TREES AND PEOPLE

TREES AND PEOPLE

FORESTLAND ECOSYSTEMS AND OUR FUTURE

RICHARD N. JORDAN

Regnery Publishing, Inc.
Washington, D.C.

Library of Congress Cataloging-in-Publication Data

Jordan, Richard N., 1930-
Trees and people forestland, ecosystems, and our future / Richard N. Jordan.
P.   cm.
Includes bibliographical references (p.) and Index.
ISBN 0-89526-483-8
1. Forest conservation—United States. 2. Forest ecology—United States. 3. Forests and forestry—Environmental aspects—United States. 4. Forest products industry—United States. 5. Forest policy—United States.   I. Title.
SD412.J67 1994 333.75'16'0973—dc20   94-27423
                                                CIP

Published in the United States by
Regnery Publishing, Inc.
An Eagle Publishing Company
422 First St., SE, Suite 300
Washington, DC 20003

Distributed to the trade by
National Book Network
4720-A Boston Way
Lanham, MD 20706

Printed on acid-free paper.

Manufactured in the United States of America

10 9 8 7 6 5 4 3 2 1

Books are available in quantity for promotional or premium use. Write to Director of Special Sales, Regnery Publishing, Inc., 422 First Street, SE, Suite 300, Washington, DC 20003, for information on discounts and terms or call (202) 546-5005.

To Donna

CONTENTS

LIST OF EXHIBITS

LIST OF PHOTOS

Credits:
    Author - Photos 1-12, 26, 27, 38, 41
    Mike and Rita Lowell - Photos 13-25, 29, 30, 34-36, 39, 40, 42-45
    RZA AGRA Inc.- Photos 31-33
    Westvaco Corporation - Photo 37
    Weyerhaeuser Archives - Photo 28

ACKNOWLEDGEMENTS

The little beads of perspiration appearing on my forehead in the mid-1940s in the forestlands of the state of Washington were the real beginning of the story of *Trees and People*, though serious research did not begin until the late 1980s. The most important thank you for assistance is for Donna, my wife and partner for forty years. Without her patience, encouragement, and word processing skills, the book project would never have stayed on track. Our research covered forestlands in Europe, all through the United States, and Western Canada. People who love trees and forestlands are a special breed and we owe much to many for their generous support in pulling this story together.

At the top of the list are peer reviewers of the entire manuscript: Ted Bauer, Medford, OR; Art Greeley, Bethesda, MD; Bill Hagenstein, Portland OR; Chuck Hoonan, Moraga, CA; Art Smyth, Alexandria, VA; and Scott and Adelaide Wallinger, Darien, CT. Our friend and historian, writer, author and editor, Chuck Twining, Federal Way, WA, gave the work a large boost.

There were numerous contributions of expertise and reviewers with a vested interest in various sections of the manuscript including: Herb Malany and Dave VanDeGraff, Boise, ID; Tom Holmes and Wayne Green, Port McNeill, BC; Mary Bullwinkel, Scotia, CA; Dean Solinsky, Trinidad, CA; Dan Kincaid, Troy, NC; Bill Baughman and Norm Spell, Summerville, SC; Donna Brown, Rich Long, John McMahon, and Megan Moholt, Federal Way, WA; Royce Cornelius, Enumclaw, WA; Curt Copenhagen, Dick Ford, and Rob Quoidbach, Longview, WA; Jim Clarke, North Bend, OR; John Aram, Sam Brown, Phil Hayes, Dan Koch, and Tom Murray, Jr., Tacoma, WA; Cathy Baldwin and John Blackwell, Portland, OR; Joe Thomas, Verona, NJ; Gene Cartledge, Wayne, NJ; Rich Lewis, Gary Moll, and Al Sample, Washington, DC; Steve Conger, Weldon, NC; Nat Giustina, Eugene, OR; Dick Wendt, Klamath Falls, OR; Duane Vaagen, Coleville, WA; and Bill Berry, Stamford, CT.

Many educators, historians (both professional and amateur), long-time observers, and true conservationists were helpful including: Dean Larry Tombaugh and Bob Slocum, Raleigh, NC; Dean David Thorud and Dave Leland, Seattle, WA; Dean Gregory Brown, Otis Hall, and Jack Muench, Blacksburg, VA; John Gordon, New Haven, CT; Dean Garry Brewer, Ann Arbor, MI; Mary Matthews and Pete Steen, Durham, NC; Mike MacMurray, Bend, OR; Jim Rathbun and Bruce Vincent, Libby, MT; Arnie Bolle and Dale Burke, Missoula, MT; Joe and Pat Slate, Silver Spring, MD; Don Willeke, Minneapolis, MN; Dick Sorensen, Wycombe, PA; Perry Hagenstein, Wayland, MA; Zach Stewart, San Francisco, CA; Lynn Day, Grosse Pointe, MI; Bud Tomascheski, Eureka, CA.

Professional forestry organizations and government agency people, particularly the Forest Service family, deserve a huge thank you. Rather than list Forest Service names, from the Auditor's Building in Washington, DC, throughout the field offices—there would be hundreds—it is best for each of you to know that Donna and I are truly grateful for your unending professional and friendly assistance; Forestdirektor, Dr. Gunter Braun, Taufkirken, Germany; Forestdirektor, Dr. Georg Meister, Bad Reichenhall, Germany; Forestdirektor Hermann Schafer, Karlsbad, Germany; Ministerialdirektor Otto Bauer and Oberamtsrat Dietrich Zerneke, Munchen, Germany; Bob Izlar, Norcross, GA; Ed Cone, Forest Park, GA; Phil Schenck, Avon, CT; Peter Grieves, Newberry, MI; Keith Olson, Kalispell, MT; Greg Smith, Bethesda, MD; Keith Argow, Vienna, VA; Ron Sheay, Trenton, NJ; Bill Pickell, Olympia, WA; Bill Dennison, Sacramento, CA; and Rick Brown, Arnold, CA.

Most of the photos are the work of Mike Lowell who gave unstintingly of his time in the field and in the darkroom. Mike's contribution adds an invaluable dimension to the story of *Trees and People*. Thank you Mike, Rita, and Sarah.

Last but not least are the people of Regnery Publishing. Indeed it was a glorious day when Al Regnery became interested in our story. Only then did we gain the insights and the assistance necessary to complete the manuscript. And, of course, we are indebted to Jennifer Reist for moving our manuscript expeditiously through the process to the final printing —always with a smile. Our editor, Patricia B. Bozell, is an extraordinary talent—beyond words. With loving kindness and a warmth which makes one bask in the final result, she strips away superfluous words, phrases, and even a chapter to emphasize the essential elements of our story.

The Forest Conference convened in Portland, Oregon, on 2 April 1993 by President Clinton was an extraordinary gathering of a broad cross-section of people concerned with forestland conservation. The general public had not witnessed such a gathering since the American Forest Congress in 1905 which received the personal attention of President Theodore Roosevelt.

President Clinton called for all participants to speak and listen from the heart. He deserves much credit for his efforts to find common ground in disputes that have arisen from our shared concern for trees and forestlands.

My story, *Trees and People*, also seeks common ground. Part I gives the background on the general subject of forestland ecosystems and helps the reader to unravel much of the confusion surrounding the many contentious issues. There is also an illustrated, in-depth analysis of stewardship.

Part II attempts to interest the lay person in forestland conservation and management in both our urban and rural forestlands. It also outlines a new paradigm for forestland conservation in the twenty-first century and suggests that Forest Products Industry CEOs seize this unique opportunity and take up the baton of visionary leaders.

Part III traces the history of forestland conservation and pays tribute to the inspirational leadership of a small number of pioneer conservationists. It also gives the lay person an opportunity to "go back to school," to immerse himself or herself in a brief but useful history of forestland conservation and professional forestry.

Before, during, and after President Clinton's Forest Conference, heated rhetoric reached new heights. An ordinary citizen would be bewildered by the bombast. Whom to believe? What to believe?

We know the United States has enacted a series of environmental laws since World War II, which has moved us in the right direction in caring for our forestland ecosystems. But in recent years, critics complain that the spirit of these laws has been abused by those who seem more interested in politics and fund raising than in the environment.

Criticism has also been heaped on the Forest Products Industry, even though it has been successfully renewing our forestlands for the past fifty years. Industry leaders, of course, had no choice but to practice sound forestry; they faced extinction had they continued the three-hundred-year-old practice of exploiting our forestlands as if they were inexhaustible.

A good deal of forestland conservation research was necessary to write this story. I was compelled to retrace my steps over the past forty years to make sure today's confusing rhetoric had not separated me from my senses, from what I understood to be guardianship of our forestland ecosystems during my own career.

Professional resource managers will understand the story of *Trees and People*. But that is not enough. The story will miss its purpose if it does not attract the attention of lay people and stimulate budding citizen conservationists.

Though neither I nor my partner, Donna Jordan (see photo 1), is a professional author or photographer, this is our personal story and it is offered to lay people as well as professional resource managers in the hope that it will shed fresh light on a contentious issue—a search, if you will, for common ground, a means of sharing our love for our renewable forestland ecosystems.

PART I

A PRESENT-DAY VIEW OF FORESTLAND ECOSYSTEMS

1

FORESTLAND ECOSYSTEMS:
WHERE TO FROM HERE

OUR LIVES ARE inextricably tied to trees and forestlands, spiritually, aesthetically, and because of the benefits they afford. The union between trees and people took place before the dawn of civilization. With few exceptions, our lives are entwined with trees and forestland ecosystems from the cradle to the grave.

Thus when trees and forestlands are endangered or destroyed by humankind or Mother Nature, our emotional reactions are deep-seated and intense. It is time we share our concern—speak of it in a more positive way. That is what this story of trees and people is all about.

We begin with the controversial and confusing issue of "old-growth forests."

Few people, scientists or lay persons, can agree on what old growth is. The general consensus, so often depicted by the media, is that environmental activists occupy the high ground. They define old growth as those trees that existed before the colonists arrived. In their view, old growth should be left in its wild and natural state in order for society to benefit fully from its presence. They therefore call for its preservation in particular, and for a regulatory approach to environmental activism in general.

The Forest Products Industry, whose primary mission is to satisfy consumer demand, looks upon forestland ecosystems, including old-growth trees, as renewable resources which require careful management and common sense stewardship. Without these renewable resources, the industry would cease to function. Unfortunately, the media have successfully depicted the old-growth issue as "owl vs. logger." This is too simplistic.

When the preservationist ethic and common sense stewardship differ over the ways in which trees and forestlands are managed, public reactions can become heated. Good citizens do care—and care strongly—about the environment in general and about trees and forestlands in particular.

Lacking information, or because of an understandable disinterest in the complexities of silvicultural sciences, important elements of the media seem to believe that the general public will only understand a simplistic approach—and in the broad sense this may be so. But the public has demonstrated that it expects responsible leaders to go beyond emotion and the simplistic approach when it comes to making decisions that affect the nation and the environment.

The tide of emotion surrounding the clash between the preservation ethic and the survival of the Forest Products Industry involves far more than owls and loggers. What has evolved in the maze of legislative initiatives and lawsuits is regulatory gridlock. There are so many conflicting laws that Washington insiders refer to the gridlock as "goal incongruity." Whatever the name, the reality is that professional resource managers can no longer move in any direction without breaking one or more laws.

On the 1992 campaign trail in the Northwest, candidate Bill Clinton quickly recognized the confusion brought about by regulatory gridlock. He promised the people in the area that the problem would receive high priority in his administration.

True to his word, President Clinton convened a Forest Conference in Portland, Oregon, on 2 April 1993, a bold undertaking for which he deserves much credit. The conference riveted the attention of the media and the general public on the contentious issues surrounding our forestland ecosystems.

In part, this conference was reminiscent of the American Forest Congress of 1905 which was supported by President Theodore

Roosevelt and which achieved lasting benefits to society regarding conservation of our forestland resources.

At his conference, President Clinton set the tone and sought to pull together the economic and environmental interests of all participants and to reach a balanced solution. He urged participants to speak and listen from the heart as he obviously did throughout the long day. It remains to be seen how and when the fruits of his Forest Conference will ripen.

CONFUSION REIGNS

President Clinton presented his findings and recommendations for a resolution of the regulatory gridlock on 1 July but they have so far only evoked loud complaints from all sides. Again, the approach seems to be too simplistic; it does not come to grips with the facts.

An example of why regulatory gridlock is confusing is the job issue. The administration's plan, referred to as "Option Nine," was presented as a balanced solution. It stated that only six thousand direct jobs would be terminated. But the Forest Products Industry estimates that 85,000 direct and indirect jobs will be lost.[1]

The job issue is understandably a critical issue, especially if it is your job. And it tends to obscure the more critical issue of the availability of raw material from our federal forestlands for the benefit of society at large.

Forestlands in Northern California, Oregon, and Washington are a classic example of America the Beautiful. They show why in the United States we can count our blessings regarding renewable forestland ecosystems. A combination of soils, climate, and species have created pristine forests which are found nowhere else in the world.

These majestic forestlands are also the most productive in the world. In the private sector, this is a tremendous economic advantage. Not so the public sector, which has dissipated the advantage through conflicting legislation that has resulted in regulatory paralysis.

An essential element of the preservation ethic is the desire to protect wildlife and habitat. Option Nine endorses this end by greatly expanding federal reserves. These "designated" reserves, which will be detailed in due course, include nonforested areas as well as some of the nation's most productive forestlands. But since 1990, the ever-expanding reserves to protect wildlife in the public sector have cut back harvest levels dramatically.

For the Northwest, the administration's plans call for reducing the level of annual harvest to 1.2 billion board feet (BBF) for the next ten years. This compares to an average of some five BBF in the 1980s. Advocates for the protection of wildlife and biodiversity are angry because they claim this reduced level does not provide adequate protection. The rhetoric on both sides has become thick and heavy.

By and large, media coverage does little to enhance public understanding of the issues. For example, one reporter recently referred to the "Northwest's long-running fight over its old-growth forests." True, old growth has held the limelight. But old growth is no longer the primary raw material from federal lands.

Moreover, most of the remaining old growth is protected in designated reserves such as national parks and Wilderness. The old growth that is available for harvest is heavy to salvage in contrast to first quality raw material found in healthy, thriving, managed forestlands. Overmature forest stands are "sick" ecosystems; they are rife with rot, insects, and disease and have suffered wind and wildfire damage. In fact, salvage sales, often blocked in the courts by antiforestry advocates, have proved to be the most successful method of nurturing an overmature, sick forest stand back to health.

In another example, another reporter said we were facing a battle between "environmentalists" and the "timber industry." Does this mean that environmentalists are not concerned about the economic well-being of society, and that "timber interests" are not concerned about the environment? When it comes to renewable resources and our environment, in reality, the choices are not either-or. There is a basis for common ground; and people everywhere can come to an agreement about how best to care for our renewable forestland ecosystems.

WHERE DO WE GO FROM HERE?

President Clinton's Forest Conference provided an excellent public forum and a golden opportunity for all parties to seek common ground. If we are to overcome regulatory gridlock, we must work from the heart. Fortunately, we have a firm base from which to start: Who does not love trees and forestlands?

There is, however, an overriding problem. Too often, we are in the dark about critical forestland ecosystem issues. If we are to play a positive role in finding common ground, we must become informed and then become active citizen conservationists.

When we are willing to go beyond simplistic rhetoric, the ecological evidence of sensible stewardship is not in question. It is this: the renewable resources that make up our environment have inherent characteristics; they are resilient, dynamic, changeable, and changing. This renewal process for air, water, soil, flora (plants), fauna (animals), and people—not to mention raw material for our essential daily needs—has been going on for millennia.

The renewal process lends itself nicely to the new concept of ecosystem management whereby both quality and quantity of these resources are enhanced. Moreover, inherent characteristics and ecosystem management are the best means to make things right when environmental problems do occur from abuse, misuse, natural disasters, or human accidents.

In short, ecosystem management, coupled with the resilient nature of our renewable resources, offers plenty of common ground, unlike the simplistic perspective of the antiforestry advocates, who embrace a finely focused and emotionally based agenda. Their vituperations make one believe that they fear common ground would undercut their financial backing and fund raising. They lack any reliable scientific data to support their narrow agenda which is based on the perception that our renewable resources are essentially nonrenewable, are subject to permanent degradation, and are always damaged, if not destroyed, by professional management. This perception simply doesn't square with reality.

But ample evidence shows that this simplistic view is widely held by the general public. Part of the reason is that three out of four people live in urban areas and, never having tasted country living, cannot separate myth from reality. The vast majority are unaware that back in 1875 citizens gave life to a conservation ethic for trees and forestlands.

This ethic and ecosystem management pertains to all of us. We must realize that city trees deserve our care every bit as much as trees in remote rural areas. As for the wide open spaces, our renewable forestland ecosystems are worth caring for whether they are found in designated reserves or in commercial forestlands.

Trees, wherever they are found, should not be left on their own. They are a part of our being. To ignore the care of our forestland ecosystems makes no more sense than to ignore the health care of people.

FORESTLAND CONSERVATION: HISTORICAL PRECEDENTS

History provides a precedent for a common effort by citizen conservationists, professional resource managers, and members of Congress. In 1875, Dr. John A. Warder and a small group of horticulturists, nurserymen, botanists, and ordinary interested citizens, twenty-five in all, breathed life into a conservation ethic for trees and forestlands.

At this meeting, Dr. Warder was elected president of the American Forestry Association, the first citizen conservation organization in the United States. At the time, there was not a single forester or professional organization concerned with forestland conservation. The association was soon joined by professional foresters (Bernhard Farnow was the first in 1876) and together these conservation pioneers worked with Congress on a series of legislative initiatives which eventuated in the renewal of our forestlands after World War II.

The first fifty years of forestland conservation were noted for enabling legislation. The past fifty years, in contrast, have stood out for the predominance of restrictive legislation leading to regulatory gridlock.

The beneficial legislation launched cooperative forestry pro-grams throughout all forested regions and gave rise to the concept of multiple-use. After World War II, a race developed between con-sumer demand and high-tech industrial forestry. So far producers have managed to stay ahead of demand by renewing our forests rather than eliminating them, but this could change, and soon, unless we overcome regulatory gridlock.

Since World War II, the Forst Products Industry, as document-ed in Chapter 3, has been on an ecologically sound footing; it is, for example, solar driven as is no other industry of its size. Its resource base, moreover, is renewable in every sense of the word. Abundant renewable raw material, after all, is the key to satisfying consumer demand. Yet, incredibly, the industry has received no gratitude from the general public. Rather, people seem to prefer to believe the steady stream of media misinformation asserting that the Forest Products Industry, the U.S. Forest Service, and professional foresters have little interest in the very forestland resources that provide their livelihood.

This misunderstanding by lay people defies common sense. They unthinkingly accept the proposition that our renewable resources are not renewable and that, for obscure reasons, they are being destroyed by professional management. These resources, in fact, have renewed themselves after natural disasters for eons, includ-ing ice ages, wildfires, hurricanes, and volcanic eruptions. Immedi-ately after the eruption of Mount St. Helens in 1980, media reports led us to believe that we would not live long enough ever again to see wildlife or plant life within the blast zone.

MOUNT ST. HELENS: DISASTER AND RENEWAL

The Forest Service has long been aware of the magic of Spirit Lake country surrounding Mount St. Helens, which was designated a "Primitive" area (a forerunner of "Wilderness") before World War II. But this wondrous natural environment, complete with road access, attracted visitors from every state in the Union and foreign lands as well. Picture a three-and-a-half mile trail around the east side of Spirit Lake leading to campgrounds, a cabin resort, "Y" camps, and beyond to the high ridges and high lakes country north of Mount St. Helens.

Touch with me the magic of an early morning stroll along this trail before the eruption. Mists like spirits rising from the lake. A whiff of bacon cooking over an open campfire. Smoke curling from the chimneys of a quiet cabin resort nestled up close to Harmony Falls. The last mile ascended nearly 100 feet above the shoreline. To walk the trail was like walking on the softest carpet. The evergreen-scented air was just short of intoxicating. The towering Douglas-firs, noble firs, and hemlocks on both sides of the trail, clinging to the steep ground were so tall that their tops, along with the sky, were obscured in a dense green canopy.

And then, disaster! Nearly all of the giant old-growth trees were incinerated. They disappeared in a cloud of dust. Only a few remained as floating debris on the surface of a new lake. The trail is gone now, along with the trees and the soil. Bare cliffs in multicolors from the heat of the blast stretch from the waterline to the top of the ridge. On the steep ground, no trees, and mostly no soil; for miles, nothing but a sterile landscape—barren ash plains—no sign of life.

And yet God, in "mysterious ways His wonders to perform," was immediately on the scene, as were scientists of every discipline. As soon as the ground cooled and Spirit Lake quit boiling, they swarmed into the blast zone where many have remained with long-range research projects. The Volcanic National Monument (110,000 acres) is probably the most researched piece of ground in the world and is still yielding some of the best scientific data ever accumulated on forestland ecosystems.

There has been a return to life of over a thousand identified species within what was thought to be a sterile environment. These findings, among other things, are in stark contrast to antiforestry rhetoric which would have us believe species can only be saved if their habitat is protected.

The revival of the blast zone provides another case history of the natural resilience of forestland ecosystems. Also within the blast zone, we can witness ecosystem management by professional foresters and other members of Weyerhaeuser's research team on their St. Helens Tree Farm. This concept should be implemented wherever trees and forestlands are found. Ecosystem management takes full advantage of the inherent characteristics of our renewable resources and assures for society their countless benefits.

In reality, the interests of our society at large are best served when we recognize that our forestland ecosystems are resilient and dynamic and renew themselves even without any assist. But as consumer demands continue to grow, we will need all our science, technology, and skills to enhance nature. We cannot afford to defer to the long natural cycles of several hundred years advised by antiforestry advocates.

To depend totally on "nature knows best" as suggested by these advocates is folly, at times more serious than folly. Al Gore, in his book *Earth in the Balance*, touts nature above all else. He says:

> The Pacific yew can be cut down and processed to produce a potent chemical, taxol, which offers some promise of curing certain forms of lung, breast, and ovarian cancer in patients who would otherwise quickly die. It seems an easy choice—sacrifice the tree for a human life—until one learns that three trees must be destroyed for each patient treated, that only specimens more than a hundred years old contain the potent chemical in their bark, and that there are very few of these yews remaining on earth.[2]

This last sentence is antiforestry rhetoric at its worst. What does Mr. Gore mean by "sacrifice the tree for a human life"? When one enjoys a glass of wine, is the vineyard "sacrificed"? When one enjoys an ear of corn, is the cornfield "sacrificed"? And what could he possibly mean by "very few" yews remaining on earth? The natural range for this species runs from Southeast Alaska along the Pacific Coast to Monterey Bay, California, and then east into the Sierra Nevada Mountains. In the north the range extends eastward through British Columbia and Washington to the west slopes of the

Rocky Mountains in Montana and Idaho. Yew trees are found from sea level to 8,000 feet. There are many millions of yew trees throughout this range, not to mention the millions of mature yew trees protected by law in designated reserves.[3]

And what about humankind? Without raw material, mills could not operate, homes could not be built and furnished, schools and offices could not be supplied, and Mr. Gore's book could not have been published. And, of course, the majority of wage earners in forested regions would be out in the cold.

In short, products essential to our standard of living require the harvesting of trees. Ironically, few people understand that harvesting is the first step of forestland renewal, whether done by nature or managed by professionals.

In the last decade of this century and into the next, concerns for the environment will be uppermost in the minds of citizens everywhere. Unfortunately, the media often suffuse these concerns with negative rhetoric. Isn't it time to renew the conservation ethic for trees and forestlands begun by a handful of citizens in 1875? We have much more going for us as we close out the twentieth century.

We have more professional foresters and schools of forestry than any other country. We are blessed with climate, soils, and species which give us renewable resources like no other people on earth. Our forestry science, technology, and skills are the standard of comparison worldwide. And we have dedicated and competent research teams and employees in our forestlands all across the country.

We thus have the means to find common ground by sharing, in a positive sense, our concern for trees and forestlands. To do this we must become informed about ecosystem issues. We can then pull together and assure the future well-being of our children and grandchildren.

2

FORESTLAND ECOSYSTEMS:
PANDORA'S BOX

THE MANY CONTENTIOUS issues surrounding our forestland ecosystems have become a gigantic Pandora's Box, and there is no way for professional foresters to know what will pop out from one day to the next. For a better understanding of this dilemma, a thumbnail sketch of forestland conservation history is appropriate. A review will also give the reader a perspective, detailed in Chapter 8, concerning enabling legislation during the first fifty years of forestland conservation as contrasted to the restrictive legislation of the past fifty years.

History reveals that Congress had little interest in conservation during the first hundred years of our Republic. Then, for the next fifty years, thanks to a small group of citizen conservationists and professional foresters, Congress passed enabling legislation for forestland conservation and management.

The Forest Reserve Act of 1891 was the first breakthrough. Next came the Forest Management Act of 1897, which was even more important as it established the fundamental policy for managing the National Forests. In addition to protecting forestlands from wildfire and protecting watersheds, this law declared that the forestland reserves were created "to furnish a continuous supply of timber for the use and necessities of citizens of the United States."

The Forest Reserve Act of 1891 was celebrated as the centennial of the National Forest System in 1991. But actually, the system as we know it today was not founded until the Transfer Act of 1905. This act was described in President Theodore Roosevelt's autobiography:

> In 1905, the obvious foolishness of continuing to sepa-rate the foresters and the forests, reenforced by the action of the First National Forest Congress, held in Washington, brought about the Act of February 1905, which transferred the National Forests from the care of the Interior Department to the Department of Agricul-ture, and resulted in the creation of the present United States Forest Service.[1]

Until the transfer was made, the federal forestlands were under the jurisdiction of the Department of the Interior while near-ly all of the nation's professional foresters were employed by the Bureau of Forestry in the Department of Agriculture.

The Forest Management Act of 1897 was strengthened by the Weeks Law of 1911, and strengthened further by the Clarke-McNary Act in 1924. This series of beneficial legislative acts enhanced the view that our nation's forestlands were a renewable resource. As a result, for the first time Forest Products Industry operators were given an opportunity to embrace conservation and still respond to the needs of citizens. But up through the first half of the twentieth century, pos-itive case histories of forestland renewal were few and far between.

This changed after World War II. Since then, the remarkable progress in renewal of our forestland ecosystems has been accompanied by an equally remarkable series of conservation and environmental laws. These were never intended to restrict forestland conservation and management, but they did in fact result in regulatory gridlock.

A brief review of the key legislative benchmarks governing our renewable forestland resources is necessary in order to understand better the Pandora's Box dilemma.

MULTIPLE-USE SUSTAINED YIELD ACT, 1960

Signed into law on 12 June 1960, this act was described as "congressional codification of 75 years of Forest Service tradition and policy." Further, "The law was considered to be supplemental to the 1897 Forest Management Act."[2]

Multiple-use was based on equal consideration of the various uses of the National Forests, while also presuming that combinations of these resources were to be managed in a way "that will best meet the needs of the American people." Productivity was not to be impaired and management was not necessarily to be measured in economic terms.

The resources of the National Forests would be managed on a sustained yield basis, defined as "maintenance in perpetuity of high level. . . output of the various renewable resources." That is, the act "directed the Forest Service to give equal consideration to outdoor recreation, range, timber, water, and wildlife and fish resources and to manage them on the basis of sustained yield."

WILDERNESS ACT, 1964

Next was the Wilderness Act, signed into law by President Lyndon B. Johnson on 3 September 1964. This act gave legal recognition to the National Wilderness Preservation System, which comprised fifty-four units of wilderness and wild areas, totalling 14 million acres, already established in the National Forest System.

The act also called for reviewing within ten years another thirty-four units of primitive areas which comprised 5.4 million acres. Wilderness was defined as "an area where the earth and its community of life are untrammeled by man, where man himself is a visitor who does not remain."

ENDANGERED SPECIES PRESERVATION ACT, 1966
ENDANGERED SPECIES CONSERVATION ACT, 1969
ENDANGERED SPECIES ACT, 1973

The Endangered Species Preservation Act, signed into law on 15 October 1966, was the "first act to protect endangered species." Under its terms, the secretary of the interior was directed to "carry out a program of conserving, protecting, restoring, and propagating selected species of native fish and wildlife."

Next came the Endangered Species Conservation Act of 5

December 1969. This act directed the secretary of the interior to develop a list of threatened wildlife and expanded the fish and wildlife categories "to include any wild mammal, fish, wild bird, amphibian, reptile, mollusk, or crustacean."

And finally, the Endangered Species Act became law on 28 December 1973. This act "provided for the protection of ecosystems of endangered species, expanded definition of wildlife, established distinction of endangered and threatened species and the concept of critical habitat."

NATIONAL ENVIRONMENTAL POLICY ACT, 1969

Accompanying the Endangered Species Acts was the National Environmental Policy Act (NEPA), which was signed into law by President Richard Nixon on 1 January 1970. The *Encyclopedia of American Forest and Conservation History* quotes President Nixon as follows:

> It is particularly fitting that my first official act of the new decade is to approve the National Environmental Policy Act—the 1970s absolutely must be the years when America pays its debt to the past by reclaiming the purity of the air, its waters and our living environment. It is literally now or never.

The results were far-reaching, the Encyclopedia continued:

> NEPA's requirement for environmental impact statements proved to be a powerful tool, or weapon, depending upon one's point of view. . . environmental groups skillfully manipulated the impact statement requirement to cause or to threaten costly delays in federal projects, as a means to attain changes favorable to their point of view.

FOREST AND RANGELAND RENEWABLE
RESOURCES PLANNING ACT, 1974

This legislation, signed into law on 17 August 1974, was referred to initially as the Humphrey-Rarick Act but is now more commonly known as RPA. It directed the secretary of agriculture to assess all lands and prepare a management program. The first assessment was to be completed by 31 December 1975 with updates in 1979 and every ten years thereafter.

The most recent assessment provides for "an analysis of present and anticipated uses, demand for, and supply of the renewable resources of forest, range, and other associated lands with consideration of the international resource situation, and an emphasis on pertinent supply, demand and price relationship trends."

NATIONAL FOREST MANAGEMENT ACT, 1976

The RPA of 1974 was amended by the National Forest Management Act (NFMA), which became law on 22 October 1976. This new legislation was said to provide a comprehensive blueprint for the management of National Forests. It reaffirmed the intent of RPA and assured public involvement in the assessment and planning process.

NFMA endorsed the concept of multiple-use and expanded the definition to include maintenance of a diversity of plant and animal species. It recognized clearcutting as an acceptable silvicultural practice. And it also gave statutory status to the National Forest System, which had existed since 1891 under a series of presidential proclamations.

U.S. FOREST SERVICE: UNTENABLE POSITION

Today, the impact of so much restrictive legislation has placed the Forest Service in a virtual straightjacket. The 1,200 page 1993 USDA Forest Service document entitled "The Principal Laws Relating to Forest Service Activities" outlines 226 laws.[3] And they—principal laws—are often at cross-purposes. That is, regarding forestland conservation and management, professional resource managers cannot move in any direction without breaking one or more laws

Criticism is heaped upon the Forest Service for being the "messenger" that heralds this series of laws. But maligning the Forest Service serves no useful purpose. The problem rests with the laws as they are presently structured, not with the Forest Service personnel or the courts that have no recourse but to implement the laws.

THE PHANTOM FOREST PRODUCTS INDUSTRY

In spite of these restrictive laws, the Forest Products Industry has been able to achieve a miracle by supplying the essential daily needs of our society. And it is no small miracle. Think for a moment. Since World War II our population has nearly doubled, from 134 million to over 260 million. The increase alone is more than the combined population of England and France. Yet, we have never had a shortage of our daily needs from our forestlands.

The miracle began when industry leaders realized that in order to survive they had to start renewing our forestland ecosystems rather than eliminating them as they had for over three hundred years. Cooperative forestry programs between federal and state agencies and private forestland owners contributed greatly to the miracle. It *is* possible to pull together.

This story is worth telling, but industry leaders have not effectively communicated to the public what they have accomplished. For the past thirty years, various antiforestry advocacy groups within the environmental community have filled the vacuum with criticism of what they deemed is unacceptable behavior by industry. In doing so they ignore ecological science and the resilient and dynamic nature of our renewable resources. Incredibly, these critics have convinced the general public that industry is destroying rather than renewing our forestland ecosystems and that damage control by industry has so far not solved the problem.

REPORT OF THE FOREST SERVICE FISCAL YEAR 1992

All the while, Forest Service crews dig in each day and strive to carry out their mission: caring for the land and serving people. How they do this is shown in considerable detail in the *Annual Report of the Forest Service*, Fiscal Year 1992.

The report highlights programs that advance the four strategic themes of the RPA:[4]

- Recreation, wildlife, and fisheries enhancement

- Environmentally acceptable commodity production

- Improving scientific knowledge about natural resources

- Responding to global resource issues.

The report also gives the lay person a better understanding of the issues surrounding forestland conservation and management.

Forestland Ownership

The total landbase of the United States is 2.271 billion acres.[5] About one-third, 728 million acres, is forestland. Of this, 213 million acres are noncommercial forestland. The remaining 483.3 million acres are commercial forestland. These forest stands annually produce 20 cubic feet/acre or more of wood fiber. Commercial forestland ownership is as follows:

Table 1

COMMERCIAL FORESTLAND AREA BY OWNERSHIP, 1987

	Million Acres		
National Forest	85.2		
BLM	5.8		
Other	6.0		
Total Federal		97.0	
Total State		26.7	
Total County & Municipal		7.0	
Total Indian		5.6	
TOTAL PUBLIC		136.3	(28%)
Private Farmers		97.0	
Private Other Nonindustrial		179.4	
Total Private Nonindustrial			276.4	(57%)
Total Private Industrial			70.6	(15%)
TOTAL FORESTLAND			483.3	(100%)

Source: Forest Service, USDA, PNW-RB-168, 1989

Designated Reserves: A Cover-up

Note that the land base for commercial forestland in Table 1 does not include 34.5 million acres of "designated reserves." These reserves are fully stocked commercial forestlands that have been withdrawn from utilization by statute or administrative regulation.

The removal of these reserves from our renewable forestland resource data base as stipulated by Congress amounts to tampering with the facts and impedes the ability of the Forest Service to carry out its charter.

The designated reserves include forestlands in Wilderness Areas, National Parks, and many other set-asides. Moreover, the laws affecting these reserves provide for little or no forestland conservation and management.

What is the magnitude of these reserves? If 34.5 million acres were included in the commercial forestland base, we would have 517.8 million acres. The reserves, then, are 7 percent of the total, an enormous area, beyond the grasp of most of us. They equal the combined size of the states of New Jersey and New York.

These congressional initiatives are disturbing. Even more disturbing is the recent acceleration of the trend. For example, in January 1992, 6.9 million acres of renewable forestland ecosystems were set aside as habitat for the northern spotted owl. And the movement to protect threatened and endangered species continues.

A huge expanse of renewable forestland ecosystems is being added to designated reserves and thereby becomes off limits to ecological science. Obviously, if we are not permitted access, there is no way we can pursue ecosystem management.

Commercial Softwood Forestland Ownership

The figures in Table 2 reveal the immensity of the federal monopoly of softwood resources, which are concentrated in the Far West.

Table 2

1992 COMMERCIAL SOFTWOOD FORESTLAND OWNERSHIP BY %

National Forests	47%
Other Public Lands	13%
Private Nonindustrial Lands	25%
Private Industrial Lands	15%

Source: Forest Service, USDA, 1992

Total Annual Wood Harvested

One of the reasons for showing these statistics here is to demonstrate how fortunate we are that cooperative forestry has worked as well as it has over the past fifty years—something the general public is totally unaware of. The private sector's performance can be judged by examining the following Table 3 which shows total annual wood harvested from all forestlands in the United States:

Table 3
1992 TOTAL ANNUAL WOOD HARVESTED BY %

National Forests	13%
Other Public Lands	7%
Private Nonindustrial Lands	49%
Private Industrial Lands	31%

Source: Forest Service, USDA, 1992

Note that half of the available inventory of softwood resources is owned by the federal government. Yet only 13 percent of the annual wood harvest is being supplied by the National Forests. The charter of the National Forest System—"To furnish a continuous supply of timber for the use and necessities of the citizens of the United States"—is made meaningless by regulatory gridlock.

Raw Material Supply from Our National Forests

The actual annual harvest from our National Forests from 1986 through 1990 averaged 12 billion board feet (BBF). Since fiscal year 1990, the harvests were drastically reduced to 8.5 BBF in 1991, 7.3 BBF in 1992, and 5.9 BBF in fiscal year 1993. Sales in 1993 were restricted to 4.5 BBF. Option Nine (mentioned earlier) would have virtually no effect on reversing this trend.

Product Demand from Our National Forests

Let's look more closely at actual demand for forest products. The previously cited RPA analysis Report RM-199 (Table 35, Roundwood consumption in the United States, by softwoods and hardwoods, and product, specified years 1952-1986) indicates a 66 percent increase in domestic consumption, from 11.9 billion cubic feet in 1952 to 19.8 in 1986.

Against the backdrop of increasing restrictions on harvesting, the obvious question is, how in the world can supply ever meet demand? And are average citizens truly aware of how their lives would be affected if essential daily needs from our forestlands were no longer available? Should they not be uneasy about court decisions and legislative enactments that severely restrict harvesting on public lands? Are they only concerned with threatened wildlife? And not about threats to their own well-being?

Threatened and Endangered Species

The concept of the Endangered Species Act moved conservation in the right direction. It has, however, become a monstrous Pandora's Box and a major contributor to regulatory gridlock.

For example, caring for the northern spotted owl is a colossal management problem for the Forest Service. Yet, the owl is only the tip of the iceberg. Our National Forests provide habitat for 33 percent of all federally listed threatened and endangered species (243 out of a total of 728).

There are twenty-five other birds besides the northern spotted owl. In addition there are 50 fish, 27 mammals, 22 clams, 13 reptiles and amphibians, 8 insects, 3 snails, 2 crustaceans, and 92 plants. And the list is being expanded aggressively.

The Kanab ambersnail in Utah is just one example of what "aggressively" means to private property owners. Mr. Brandt Child invested $2.5 million in a five hundred acre desert oasis in 1990 where he planned to build a campground and golf course near three lakes. The Fish and Wildlife service, armed with the Endangered Species Act, stepped in and put a halt to his plans in order to protect some 200,000 snail residents of the lakes.

The *Wall Street Journal* summed up Mr. Child's dilemma in an editorial on 27 December 1993:

> When President Richard Nixon signed the species act into law in 1973 it was primarily intended to protect threatened bears, eagles and wolves. Court rulings and bureaucrats have expanded its scope to include nearly everything bigger than a microbe, including many varieties of beetles and rats. The list of endangered species now tops 800. Another 4,000 species are nominees for the list. At this rate, there will soon be only one species that will have no protection at all: Human Property Owner.

Another example of just how misunderstood the concept of the Endangered Species Act has become was revealed in a January 1992 Sustainable Forestry Workshop in Reston, Virginia. Edward O. Wilson, world-renowned biologist from Harvard University, was one of the presenters. He said that, including oceans, there are millions of species. In his book, *The Diversity of Life*, Wilson argues that "preservation of the world's biological diversity—its wealth of wild plant and animal species—is central to the maintenance of life on Earth." And added, "Fewer than 10 percent of the earth's species have scientific names making it a 'still mostly unexplored planet.'"[6]

A contentious, confusing issue? Indeed it is! Moreover, knowledgeable scientists often disagree. For example, entomologist Sue Hubbell, in her fourth book, *Broadsides from the Other Orders*, asserts, "For every pound of us there are 300 pounds of bugs," adding, "how unlikely it is that any bug species will ever become extinct."[7]

Speaking of "us," there are over 260 million people in the United States, and far too many lack the bare essentials to survive. Yet we are giving high priority and allocating billions of dollars to preserve thousands of species, many of them yet to be identified.

It has become nearly impossible to trace the costs of the Endangered Species Act. But a significant step was taken by the National Wilderness Institute in a report to the National Press Club in Washington, D.C., on 23 March 1994.

Since passage of the act in 1973, this report analyzed government cost estimates of 306 recovery plans for 388 of the 853 currently listed endangered and threatened species. Identifiable costs total $884,164,000.

The 466 other species already listed are not part of these astronomical costs, nor are the 3,996 official candidate species. Not included are the costs of an eroded tax base, reduced or terminated business activities, and jobs lost as a result of conflict.

The report concludes that the Endangered Species program is out of control and that our government has no idea of the true cost. Moreover, this in-depth analysis reveals that direct and indirect costs of the program are in the billions. Yet, in over twenty years, not a single endangered specie has been recovered and delisted.

The costs are not the real shocker in the report. Rather, it is the revelation that government scientists—quoted many times over—admit that their knowledge of listed species is close to nil! That is, too often, we do not possess adequate scientific data to decide whether or not a species should be listed.

These billions of dollars are tax dollars, yours and mine, exacted to satisfy the appetites of antiforestry advocates. And they are simultaneously raising tens of millions of tax-free dollars with emotional appeals claiming that species cannot be "saved" unless they are in designated reserves, where nurturing renewable resources is prohibited by law. When this can happen, all species, including humans, are at risk.

Can you imagine ignoring the health needs of our children, parents, grandparents, pets, and domestic animals? That makes as much sense as ignoring the health needs of our forestland ecosystems where the bulk of our renewable resources are found. Trees and forestlands are part of our being and deserve our care. But under the present interpretation of the Threatened and Endangered Species Acts, they are not getting it.

The Appeals Process

The well-intentioned conservation and environmental laws assured that the public would become involved in the forestland planning process. This is a noble concept; but the laws also raised animal and plant species to the level of humankind, thus opening the door for a wide array of initiatives that led directly to regulatory gridlock.

The forest planning process has become a bureaucratic night-mare rather than a realistic approach to "caring for the land and serving people." Since the first forest plans were finalized in 1979, more than one thousand appeals have been filed. Of these, over eight hundred have been resolved (121 in FY 1992).

The resolution of these appeals, on average, requires two years. At the end of FY 1992, 260 forestland plan appeals were still pending—320 at the end of FY 1991. The process, obviously, is time consuming and frustrating in the extreme. It is also costly to taxpayers with estimates running as high as $150 million a year.

Here again, as with the Endangered Species Act, the appeals process is a good legal concept. But the spirit of the appeals process has been abused by antiforestry advocates, for the bulk of the appeals are designed to set aside our renewable forestland ecosystems in designated reserves, thus making them inaccessible to all but a privileged few and totally inaccessible to care.

The appeals process is just another example of the troubles the Forest Service has been subjected to for years. The first hundred of these years have been fraught with every imaginable conflict. For the full-blown story, read *The U.S. Forest Service: A History* by Harold K. Steen (Seattle: University of Washington Press, 1991), an enlightened narrative by a professional historian.

NATIONAL WILDERNESS PRESERVATION SYSTEM

The total Wilderness land area within our National Forests is approximately the size of Arkansas. As of 1992, there were 34 million acres of "official" Wilderness widely scattered in 387 designated areas in thirty-six states.

The original Wilderness idea was in keeping with the concept of multiple-use. Before the Wilderness Act became law in 1964, the Forest Service had already designated 14 million acres of Wilderness, the first 5 million in 1928.

In October 1926, Chief Forester William B. Greeley (1920-28) "laid out his thoughts on wilderness in the *Service Bulletin*: Such preservation was 'highly important, if not dominant,' he thought, but it would not normally interfere with economic use of timber, forage and water."[8]

Nor has the Forest Products Industry been opposed to the concept of Wilderness. The industry *is* opposed, however, to the established pattern in Congress of setting aside highly productive renewable forestland ecosystems as Wilderness areas. Of the 34 million acres of National Forest Wilderness, over half (18 million acres) are fully stocked renewable forestland ecosystems which the industry would like to see managed as working commercial forestland.

Setting aside fully stocked forestlands flies in the face of the long-established charter of the National Forest System. When these lands are removed from the data base, the charter has in effect been altered from one of supplying the needs of all citizens to the enjoyment of the privileged few who are able to gain access.

Polls suggest that there is much confusion regarding the Wilderness concept. For example, when the question is asked, "How many times have you driven through a Wilderness this past year?" a surprising number of respondents indicate numerous times. But it is against the law to drive through a Wilderness.

Apparently the general public is also unaware of the present extent of the Wilderness System. Media coverage tends to pay small attention to the acreage already designated as Wilderness. The facts, if given the opportunity, speak for themselves.

Table 4

NATIONAL WILDERNESS PRESERVATION SYSTEM (MILLION ACRES)

Agency	No. of Units	Acreage
U. S. Forest Service	363	33.6
National Park Service	42	39.1
Fish and Wildlife Service	75	20.7
Bureau of Land Management	66	1.6
TOTAL	546	95.0

Source: Forest Service, USDA, 1992

How large an area is 95 million acres? The total land area of New Jersey, Pennsylvania, New York, Connecticut, Massachusetts, Rhode Island, Vermont, and New Hampshire add up to only 85 million acres, considerably less than what now comprises the nation's Wilderness System.

The original intent of the National Park and Wilderness systems called for Congress to recognize unique land areas by virtue of their ecology, geology, geothermal features, archeology, and unique species, such as redwoods and sequoias already set aside in California and old-growth Douglas-fir already set aside in the Pacific Northwest.

But in recent years, there has been little distinction between working commercial forestlands and Wilderness designations. As members of Congress continue to respond to pressure from antiforestry advocates, they busily write new laws that set aside more and more renewable forestland ecosystems where little or no management is provided and no harvesting allowed.

This aggressive push for Wilderness expansion impedes ecosystem management, that is, the care of all our renewable resources; air, water, soil, flora, fauna, and raw material. Nor does it aid the long-established preservation ethic which calls for protecting the unique features of our landscape.

While Wilderness, itself, is a good concept, too much of a good thing is usually bad and that is the case here; Wilderness activists cannot be satisfied in their continual press for more and more Wilderness.

Wilderness and Recreational Values

The rhetoric that influences Congress so effectively maintains that Wilderness enhances recreation while multiple-use limits it. This is simply untrue.

Recreation is given a high priority by the Forest Service. During the past four years, over 1 billion recreational visitor days (RVDs) were recorded. Clearly, the Forest Service likes to share the joy and fulfillment that can only be found in nature. Many citizens owe much to this dedicated group of conservation-minded public servants.

By contrast, the 1992 *Report of the Forest Service* informs that only 13.3 million RVDs were recorded in Forest Service Wilderness. This amounts to 4.6 percent of the recreational use in the National Forest System, the leading agency that provides recreation to the general public.

The following Table 5 gives an overall perspective of RVDs by federal agencies:

Table 5

1991 Recreation Visitor Days by Federal Agency

	Million RVDs	
Forest Service	278.8	42%
Corps of Engineers	192.2	29%
National Park Service	112.0	17%
Bureau of Land Management	45.0	7%
Bureau of Reclamation	23.4	4%
Other	1.0	1%
Total Recreation Visitor Days	652.0	100%

Source: Forest Service, USDA, 1992

The Forest Service is making every effort to expand the Wilderness experience of visitors to National Forests. These good intentions, however, seem likely to be frustrated by the Wilderness concept itself.

Let us look at the accessibility to Wilderness in more detail. Again, recreational activities throughout the Wilderness System are enjoyed by only a privileged few. But there is considerable confusion about who enjoys Wilderness and who does not—many people believe they have experienced Wilderness by driving through some remote area.

In reality, the National Wilderness Preservation System (all 95 million acres) is functionally closed to citizens. That is, out of a population of over one-quarter billion in the United States, the percentage of those financially and physically able to visit the Wilderness System is so minuscule that it is statistically inconsequential.

In order to participate in a Wilderness experience, one must be physically fit enough to get into and out of the area leaving behind only footprints, or have the financial resources to pay for the necessary assistance.

In truth, the Wilderness concept as employed on federal lands discriminates against all but a privileged few. Those denied access include youngsters unable to hike for miles but too big to be carried; senior citizens with the same limitations; the physically impaired; and the poor who can't afford the costs of even getting to the trail heads. The Wilderness Act has become a travesty, with the vast majority excluded.

In short, the Wilderness concept as it is presently implemented is discriminatory. Title III, Public Accommodations Provision of the Americans with Disabilities Act of 1990, mandates access for the disabled to public spaces such as hotels, restaurants, grocery and retail stores, banks, theaters, and concert halls, effective 26 January 1992. There are no such provisions in Wilderness Areas. The physically impaired are, in effect, denied access by federal law.

There is another aspect to the implementation of this exclusionary law. Huge areas in the Far West were closed; roads became nonroads and gates were installed. Millions of ordinary citizens who had enjoyed outdoor recreation on these public lands for generations were no longer able to do so.

This part of the Wilderness concept is difficult to explain to those who for years considered outdoor recreation part of their national heritage. They simply cannot understand why roads that gave access to campers, hunters, fishermen, hikers, and climbers have been closed, not to mention the stiff fines levied on anyone who drives a vehicle in a Wilderness Area.

Another overlooked aspect is the negative impact on wildlife of the National Wilderness Preservation System. Before these millions of acres were set aside, many recreationists, particularly in the Western states, actively participated in wildlife management. Volunteers went into remote areas and cleaned the natural water holes on a regular basis, and added others when required. Salt, another essential for healthy wildlife, was also made available.

Many of these outdoor folks were volunteers long before it became fashionable, adopting roads and trails and maintaining them for everyone's enjoyment, and setting an example regarding fires, sanitation, litter, and firearms—as well as looking out for the well-being of the young and old, experienced and tenderfoot. It is hardly surprising that these Westerners were upset when their way of life was so suddenly and radically disrupted.

Wilderness Concept: Waste:
Fantasy Mentality

There are numerous problems with the National Wilderness Preservation System. Already noted is that it is functionally closed to citizens, and that the negative effects on wildlife are significant. Additionally, many local small businesspeople dedicated to recreation have been summarily denied their livelihood; and, of course, a great many rural communities have seen their tax base eroded. Still another critical negative is the damage incurred by wildfires, insects, and disease—impossible to prevent or control without access.

But the most ominous aspect has to do with waste. In Philadelphia in 1876, when the constitution of the American Forestry Association was adopted, concern for waste in our forestlands was uppermost in the minds of the citizen conservationists. The constitution's first objective was: "The protection of the existing forestlands of the country from unnecessary waste." Until recent years, it was the conservation ethic for trees and forestlands.

But today, as Congress indiscriminately adds to designated reserves what should be working commercial forestlands, our country provides the world with further evidence that we are the most wasteful nation on earth. Will the rich heritage from our renewable forestland ecosystems be passed on to our grandchildren?

Are renewable forestland ecosystems the only issue? What about our natural resource heritage which also becomes inaccessible by law? Do we have a fantasy mentality in command that misleads our citizens and members of Congress, by arguing that our natural resources, and particularly our renewable resources, should continue to be set aside in designated reserves? If so, can it endure?

We are fortunate that our treasured forestland ecosystems are renewable. That is, they are renewable when managed. But if we leave them unmanaged, nature takes over, and nature's method is to harvest catastrophic "clearcuts" through the devastating ravages of wildfires, hurricanes, insects, disease, and old age.

RECREATION, RECREATION, AND MORE RECREATION

In our managed renewable forestland ecosystems, opportunities for recreation are unsurpassed. Yet antiforestry advocates assert that recreation is precluded unless these forestlands are isolated and left in their natural state. This is a classic example of perception vs. reality.

Recreation has been a Forest Service priority from the beginning. Greeley, in his book, *Forests and Men*, recalls the early days:

> We started with Secretary James Wilson's first commandment, "to serve the greatest good of the greatest number in the long run." In contrast to the former "reserves," the national forests were for use. And we surely put them to use—timber, forage, water, agricultural lands, minerals, recreation facilities, and wild life. We coined the phrase "multiple use" to express our zeal for the utmost public service from a section of land. We had the thrill of building Utopia and were a bit starry-eyed over it.

In respect to recreation, industrial and nonindustrial forestland owners also recognize the importance of public access to private property. Understandably in exchange for access, owners expect wise use and respect for their property. (Industrial forestland is held by large corporate firms and nonindustrial by individuals or small, often family-owned businesses.)

Our trees and forestlands across the nation have provided us with a lavish heritage, which we have and still can use for recreation as well as the enlightenment that comes from savoring the natural environment. Because of it, our lives have been enriched beyond words.

But our forestlands were not always properly nurtured. Forestland conservation and management did not emerge from the dark days of the nineteenth century until the second half of the twentieth century.

Today, the benefits of multiple-use are enjoyed by tens of millions of people through various activities—camping, boating, fishing, hiking, swimming, hunting, bird watching, photography, picnicking, canoeing, cross-country skiing, biking, climbing, horseback riding, and kayaking. Deborah Boerner-Ein in her article, "Exploring the Treasure," in the May/June 1991 issue of *American Forests* adds a few "not-so-usual possibilities" for recreation—hang-gliding, rock-climbing, spelunking, houseboating, and exploring archaeological digs.

Recreation and Managed Forestlands in Germany

Europe has long been known for its forestland recreation, and nowhere more so than in Germany. Planned public access to German forestlands holds a high priority, as is evidenced by well-maintained road and trail systems.

This was confirmed in the fall of 1989 when Donna and I, as part of our forestland conservation research, visited the Bavarian forests around Munich and Bad Reichenhall and the northern portion of the Black Forest south of Stuttgart.

We saw a remarkable sight in the company of Dr. Gunter Braun, Forstdirektor of the Sauerlach State Forest, just outside Munich—access roads criss-crossed this working forestland every 400 meters. It is one of the most highly developed road systems anywhere.

Within each grid are forest stands of a different mix of species, age class, and density. Visitors of all ages, using every means of transport, roam these healthy, thriving forest stands. There is no sense of "saving" the forest, as in the United States, but rather of enjoying its presence.

Photo 1 shows an orientation session with Donna, Dietrich Zernecke, Information Section, Bavarian Forest Service on the left, and Dr. Gunter Braun. We are in the midst of a ninety-year-old spruce forest which was selectively logged about 1980. Note the birdhouse in the nearby tree, one of some 1,500 Sauerlach wildlife homes provided by foresters and volunteers.

In photos 2 and 3 we are looking through the spruce stand

and into the late afternoon sun. In photo 2, Herr Zernecke, the keeper of the photo archives in the Munich office, is adding to his collection. Note the parklike setting of these forestlands. Nothing is wasted.

In photo 4, we see the results of a thinning operation farther along the road from the scene in the opening photo.

A thirty-eight-page booklet published by the Bavarian minister for Food, Agriculture and Forest, *The Forest, Environmental Protection in Bavaria*, notes the on-going multiple-use program and asserts, "Forests can only perform their many functions if they are correctly managed."[10] It also tells about how the Bavarian government has provided for the preservation of wilderness tracts.

The state of Bavaria has set aside 135 natural and near-natural forested areas as designated reserves, comprising a total area of 4,400 hectares or two-tenths of 1 percent of the 2.45 million hectares of Bavarian forestland. By comparison, designated reserves in the United States total 13.96 million hectares, more than twice the land area of the entire state of Bavaria.

Forests are of large importance in Bavaria. They cover more than a third of the state's land area, and citizen concern for forestlands dates back to 1368, when, as the booklet says, "Nuremberg Alderman Peter Stromer developed the first process for sowing pine, fir and spruce seeds." The earliest regulations for protecting forestlands were passed by landowners in Bavaria in 1568. Professional forestry was already well established by the end of the eighteenth century.

Based on our interviews, professional foresters in Bavaria are not certain about what a "natural" forest would be. They are clear, however, about what they wish to achieve with managed forestlands. In addition to making their forestlands accessible to everyone for recreation, they focus on other vital functions such as watershed protection, wildlife habitat, and avalanche control.

Present management objectives include reducing the percentage of evergreen as opposed to deciduous trees. Evergreens currently account for about 75 percent of the stands, with spruce nearly half the total.

As explained in the booklet, "A forest can only fulfill its diverse functions if it is correctly managed. The core of forest management is the regeneration of forests adapted to their site and tending of the forests. Regeneration provides the foundation for the future forests of our great-grandchildren."

The reader is also reminded that raw material utilization is an important part of the management process. Such utilization not only does forestland no harm, it is "indispensable to renew the forests and care for them. Only a well-tended forest can fulfill its important multiple-use tasks."

Refer to photos 5 and 6. We are with Dr. Georg Meister, Forstdirektor of the Bad Reichenhall Forest, high above the main highway just south of Bad Reichenhall (elevation about 4,000 feet). The view in photo 6 is not of a hiking trail; it is a road. By looking at a similar mountain directly across the valley (photo 6) you can see how difficult it is to construct access roads in the Bavarian Alps.

As mentioned, one of the main functions of the forest in this high country is avalanche control—protecting the highway directly below. To do that, road access is essential.

In photo 7, Donna is taking notes from Dr. Meister on avalanche control methods to stop and hold rocks and snow. These vary by site. One method is to cut ailing trees, leaving high stumps, and then lay the trees across the slope. Another is to build fences and use nets. Still another is to make tripines, large three-legged braces which are placed in a staggered formation along avalanche chutes. And last, large rakes are braced against the avalanche chute and act as a catch-all, as shown in the photo.

Photo 8 shows an uneven aged forest of spruce and fir. This is a typical view in higher elevations of Bavarian forestlands—managed, renewable forestland ecosystems in all their glory.

In photo 9, same location as photo 8, Dr. Meister, on the right, is talking with a forest products industry representative responsible for logging. These two seem to be going over details, making certain that best management practices are being followed. (That's a guess. Donna is not taking notes in German.)

Photo 10 shows the northern portion of the Black Forest below Stuttgart, just south of Karlsbad. Forstdirektor Herman Shafer is orienting us with a chart on the hood of his car. The interesting thing about this forestland is that the large trees along the road in the center and to the right that look like Douglas-firs are Douglas-firs.

German foresters obtained seeds from some of our best Douglas-fir sites in the Far West at the turn of the century. Today, Douglas-fir makes up a significant part of the harvest in the southern portion of the Black Forest, but not in the northern part.

The Black Forest is steeped in history. Forstdirektor Shafer's files detail management prescriptions from 1838. In contrast, our best case histories of renewing our forestland ecosystems go back only fifty years.

Germany is the cradle of professional forestry. In both Sweden and the United States, however, intensive forestry and the renewal of forestland ecosystems have superseded long-established management practices in Germany. Nonetheless, comparing notes with German professionals is both enlightening and uplifting. Germany is a living testimony of the beauty that results from caring for trees and forestlands.

Of course there are problems in German forestlands, but they are modest compared to ours. Catastrophic natural disasters are rare; wildfires do not rampage because fuel loads have been eliminated for centuries. Nor are insects and disease given free rein. People and nature coexist as they should everywhere.

CONFUSING RHETORIC: SCIENTISTS AND THE MEDIA

Too often, distinguished scientists—or at least scientists with impressive credentials—contribute to the confusion. We have all heard some of their contentions: When forestlands are harvested by the Forest Products Industry, they are "destroyed. . . forever!" At the same time, in the wake of natural disasters, scientists "rejoice" and are "elated" with the opportunities afforded to study forestland renewal.

Such natural disasters as the 1980 eruption of Mount St. Helens, the fires in Yellowstone National Park in 1988 and in Yosemite two years later, and Hurricane Hugo, which in 1989 blew away nearly all of the Francis Marion National Forest in South Carolina, allow us to witness the healing powers of nature.

Isn't it odd that careful harvesting can be viewed so differently from natural devastation? Clearcut technology of modest acreages occurs on individual sites, managed by professional foresters using silvicultural practices best suited for the site, species, and climate. With this technology, we also take the first step of renewal, as with nature. But the renewal process here is accelerated by science, technology, and skills developed over the past fifty years.

The new highway providing access to the visitor's center overlooking the crater of Mount St. Helens, which opened on 15 May 1993, gives visitors a stunning contrast between managed renewal and leaving renewal to nature. It shows firsthand the benefits of a managed forestland ecosystem.

And if you are not sure what ecosystem management means, you will see up close what it is all about—watershed protection, erosion control, caring for wildlife and plant habitat (many hundreds of species), providing recreational opportunities, and caring, of course, for our valuable renewable raw material. In short, giving tender, loving care to trees and forestlands.

Harvesting is really a nonissue. If we were looking at our devastated forestlands in the days of cut-out-and-get-out of the nineteenth century, the various methods of harvesting would surely be an issue, and a critical one. But this is not the nineteenth century. The real issue today is one of growing, not harvesting.

This has been true for fifty years, ever since Forest Products Industry leaders realized that exploitive methods of harvesting, which had been the pattern for over three hundred years, would lead to their extinction. The ensuing renewal of our forestland ecosystems is the most positive trend ever seen in the history of forestland conservation.

This trend is a cause for celebration in the United States; but as was amply evidenced in the 1992 Rio Earth Summit, we only hear the opposite. The general public is totally unaware of this good news. Good news, it seems, does not sell newspapers or television advertising.

Is the Media Telling the Whole Story on Forestland Stewardship?

Nowhere is there greater confusion about our renewable forestland ecosystems than in the print media and television, especially as regards harvesting. The efforts of antiforestry advocates to prevent harvesting are all too often abetted by reports in the media. These efforts border on conspiracy. Let's look at a few examples in some detail.

Time Magazine: Objective Reporting?

In the 25 June 1990 issue of *Time* magazine, the cover story, "Owl vs. Man," consisted of a seven-page feature with nine photos, a map, and a diagram. Did this feature enlighten the reader or did it foster controversy? The introductory paragraph reads:

> In Oregon's Umpqua National Forest, a lumberjack presses his snarling chain saw into the flesh of a Douglas fir that has held its place against wind and fire, rockslide and flood, for 200 years. The white pulpy fiber scatters in a plume beside him, and in 90 seconds, 4 ft. of searing steel have ripped through the thick bark, the thin film of living tissue and the growth rings spanning ages. With an excruciating groan, all 190 ft. of trunk and green spire crash to earth. When the cloud of detritus and needles settles, the ancient forest of the Pacific Northwest has retreated one more step. Tree by tree, acre by acre, it falls, and with it vanishes the habitat of innumerable creatures.

This is classic antiforestry rhetoric. Then the writer suggests that the northern spotted owl is struggling for survival. Facts, however, suggest otherwise. With a homestead of more than 6 million acres, the several thousand pairs of owls will be feathered millionaires, thanks to the allocation per pair of over two thousand acres of fully stocked forestland.

We are then informed that "man's appetite for timber threatens to consume much of the Pacific Northwest's remaining wilderness." The greater part of the wilderness of which the writer speaks is already protected by law. By "man's appetite for timber," the writer seems to be referring to our demand for forest products, as exampled by our purchase of periodicals such as *Time* magazine.

The spotted owl is described as being "a fine bird . . . but it was never really the root cause of this great conflict." Now we may be moving toward reality. At an environmental law clinic in Eugene, Oregon, in March 1988, a Sierra Club Legal Defense Fund representative acknowledged the real reason for "this great conflict":

> In fact the ultimate goal of litigation is to delay the harvest of old-growth forests so as to give Congress a chance to provide specific statutory protection for those forests. That is the long term goal. Until legislation is adopted which protects these forests, we need at least one surrogate, if you will, that will provide protection for the forests. . . .
>
> Well as the . . . as the strategy for protecting old growth matured it appeared that wildlife would offer the most fruitful hunting grounds for a surrogate that meets the three criteria. . . . Well the northern spotted owl is the wildlife species of choice to act as a surrogate for old growth protection. Thank goodness the spotted owl evolved in the Northwest for if it hadn't we'd have to genetically engineer it. . . .[11]

That at last puts matters clearly. The spotted owl serves a momentary need—delaying harvest until such time as Congress forbids it altogether.

Back to the *Time* article: The writer asserts that the dispute over the owl "merely hastened an inevitable crisis facing the Pacific Northwest. For decades, the timber industry, driven by the nation's voracious housing needs, leveled private and public land for timber with little regard for long-term consequences." Subsequently, he adds, "Clearcutting—the indiscriminate leveling of every tree in an area—has left the wilderness fragmented and scarred."

Again, we are misled. Wilderness is protected by law and no clearcutting or any other method of harvesting is allowed there. Besides, clearcutting for the past fifty years and more has been anything but "indiscriminate." It is an established, planned, and proven silvicultural practice which has been used effectively to renew forest-land ecosystems.

Moreover, if "leveling land with little regard for long-term consequences" were true, the Forest Products Industry wouldn't exist today. For nearly a half-century, billions of dollars have been invested in our forestlands, investments that will not materialize until the next century. That is a remarkable commitment of faith in our renewable forestland ecosystems.

Not so, the reporter persists, "the reinvestment is too little, too late." If Forest Products Industry operations were as poorly managed as he suggests, the many essential products for our way of life would long since have been exhausted, including such things as *Time* magazine.

The remaining pages of the article contain more of the same. At the end, a diagram depicts "What Old Growth Trees Do for the Ecosystem . . . And for the Economy." On one side, the benefits of old growth are detailed under the subheadings: "Air, Biodiversity, Soil, and Water." Renewed forestland ecosystems, of course, provide the same benefits.

As for the economic side, under the subheading "Bark & Sawdust, Lumber, Pulp and Plywood," a few essential products are listed. What is not mentioned is that these products come to us thanks to the management on a sustained-yield basis of renewable forestland ecosystems.

In *Time's* "simplistic" cover feature, full of inaccuracies and emotive rhetoric, we once again see confused the nonissue of harvesting with the real issue of growing trees.

The reporter obviously knows little about the forest products business, from harvesting raw material to manufacturing and marketing the finished products. He repeatedly castigates the "timber" industry while ignoring the vital role of the Forest Products Industry.

This approach brings to mind a *Barron's* advertisement containing a quote of Bernard Baruch: "Every person has a right to his opinion. But no one has a right to be wrong with the facts."[12] Misleading rhetoric may suit the propagandist's agenda, but is this what *Time* editors had in mind?

The answer seems to be in the affirmative, for a similar cover story appeared in the 8 July 1991 Canadian edition of *Time* under the heading, "Canada's Endangered Forests." This article offers similar distortions.

Again, the nonissue of harvesting is stressed with little mention of growing trees. The writer notes the "ugly gashes that mark the destruction of evergreen wilderness," and that "logging has left its deadly mark on Ontario, Quebec and the Atlantic provinces," as well as that "destruction in the Pacific region is more serious . . . ."

As for facts, don't look for them. We are told, for example, that "each year in Canada trees on 10 million hectares are felled for the U.S., European and Japanese markets, 90% of them by clearcutting." One million hectares would be fairly close to the mark, a sizeable error by anyone's standards.

National Geographic and More of the Same

National Geographic in its September 1990 issue covered the subject of forestland conservation and management as only it can. The article comprises thirty-one pages, nineteen photos (several of which are double page), a fold-out painting, and a two-page map.

This truly is a beautiful piece; educational, as one would expect, and replete with interesting copy and stunning photographs. Mentioned in the "On Assignment" notes are the writer's comments regarding confusion. He "found the conservation question clouded by miscommunication." Indeed!

Looked at in some detail, the copy, pictures, and accompanying comments evidence that both writer and photographer gave way to their preservationists' instincts. The result once again is to focus on the nonissue of harvesting rather than the real issue of growing trees.

The lead-in, superimposed on a magnificent two-page photo, reads: "Loggers in Washington State work in the eye of a storm over the fate of the world's greatest temperate rain forest. As the old trees fall, North Americans ask, *Will we save our own?*"

One gathers that the writer is referring to old growth; the photo, however, is a wonderful picture of harvesting in a second-growth forest. There may be a patch of old growth in the center background, largely obscured by morning mist.

What you can see are three chokersetters pulling their chokers down from a high-lead and preparing to make up a turn of recently felled logs. This is obviously not the classic old growth the writer describes in his story.

That account is preceded by six pages of photographs that pluck at the emotional heart strings. No doubt unintended, the opening photo is a perfect depiction of high-tech industrial forestry, as opposed to the old cut-and-run days. The high-lead will choke each turn of logs high up off the ground with little damage to the site as they are lifted to the landing where they will be loaded on trucks for transport to mills. High-lead logging is in contrast to tractor logging wherein a turn of logs is choked up with a cable and hauled to a designated truck loading area.

The remaining knee-deep slash in the photo will be burned and the site prepared for planting within the year. It will soon look much like the facing ridge which is a beautiful restocked young forest with a three-to-five year head start on Mother Nature.

On the opposite side of the photo, the sunlit ridge evidently was logged within the past ten years and is now covered with a healthy, lush new forest. It is a glorious example of responsible management.

The next two pages feature the acclaimed elk/old-growth photograph by Tom and Pat Leeson. And now the copy:

> Beyond the chain saw's bite, a Roosevelt elk surveys Gothic depths of an old-growth forest in Olympic National Park. Elsewhere heavy logging and hunting have eliminated two of six elk subspecies; others have been stressed almost to extinction. Only a fraction of virgin forest on public lands in the United States and Canada is wholly protected.

This is quite a statement. Where has logging eliminated elk subspecies? We do know they thrive in healthy, managed forestlands. As for "only a fraction of virgin forest" being protected, we have already noted that the designated reserves in the United States equal the combined land area of the states of New Jersey and New York.

The writer clearly has little knowledge of the Forest Products Industry. His captions refer to "timber interests. . . environmentalists and loggers. . . timbering." The issue is once more the simplistic view of "owl vs. logger."

Following the photo essay, the feature story begins, "We can't see the forest for the trees. That old saw has new teeth as logging of old-growth accelerates, while many call for a pace more in step with nature." But as the article itself amply documents, the logging of old growth is not accelerating; rather, it has nearly halted, since the bulk of remaining stands is protected by law as designated reserves.

And who are the "many calling for a new pace"? The writer must mean a handful of scientists and the most vocal antiforestry advocates. And a "pace more in step with nature"? Does the author join the "many" who prefer nature's way—the random devastation of hundreds of thousands of acres by wildfires, volcanos, and hurricanes?

At times this article departs from education and leans toward advocacy. For example, we read that "in the Pacific Northwest nine-tenths of the virgin woodland has been hauled to the mill; on the continent as a whole less than 5 percent survives."

Over the entire continent, much of the "virgin woodland" was burned to clear the land. The portion that was "hauled to the mill" at least served the needs of Americans, and is there anything wrong with that? Now and then, at the very least, writers should remember where their paper originates.

"It's a war out there," our writer continues, and "clearcuts look like battlefields." But the superb opening photograph, which shows how far we have come in industrial forestry, belies his words.

The writer goes bravely on. On redwoods: "Our beachhead against an ultimate wipe-out is a string of state parks, largely the 70-year achievement of the Save-the-Redwoods League, and Redwood National Park, midwifed by the National Geographic Society." The reality of the redwood story is that these forestlands are in better shape today than at any time during the past fifty years.

As for "what they called a 'cut and git' policy by Pacific Lumber," Pacific Lumber's record will be reviewed in the next chapter on stewardship; you be the judge.

The writer also visited Canadian forestlands. Speaking of the Franklin River Division of MacMillan Bloedel on Vancouver Island he said, "They explained how they cut and replant trees under their long-term tree-farm license from the province. I saw company mills, aircraft, and motor pools—an economic empire connoting jobs for 4,000."

So much for professional forestry; the subject is covered in thirty words. We will also review MacMillan Bloedel's forestland conservation and management in the next chapter.

In July of 1991, we motored round trip from Nanaimo on the southern end of Vancouver Island to Port McNeill at the northern end. We were looking for photo opportunities and for signs of overcutting. We found plenty of the former and very little of the latter.

Wherever harvesting had occurred, from the late nineteenth century through 1991, we found beautiful forestlands as far as the eye could see, from the first stages of renewal to closed canopies. As in the stupendous *National Geographic* fold-out of the Mount Paxton area, old-growth forestlands were conspicuous from the edges of the new forests all the way to the horizon.

On Vancouver Island, Canadians are indeed blessed with favorable soils, climate, and species. The issue of growing trees speaks for itself. And while industrial forestry has had its fits and starts in Canada as in the United States, in recent years Canadian foresters have been doing a commendable job.

The *National Geographic* writer does offer valuable insights in words and pictures on how antiforestry advocates are likely to view forestland conservation and management. The problem of misleading rhetoric stands out sharply in this article.

For example, it notes that the Pacific Northwest's Ancient Forest Alliance of eighty environmental groups has "fostered a bill in Congress that would stop all cutting of old-growth forests on federal lands."

But the owl controversy is only the beginning. These groups now intend to restrict or stop second growth harvesting in private as well as public lands.

In conclusion, the writer quotes an array of scientists. He states, "Both Jerry [Franklin] and Chris [Maser] see a smaller but endlessly sustainable forest industry. 'There must be less cutting and more consolidating of remnants into viable entities,' says Jerry. 'What the balance will be is a subject for study and negotiation. But the need to seek that balance is not negotiable.'"

Jerry leaves out a very important and necessary word—"realistic." Global demand for products from our forestlands is burgeoning, not abating. Under the circumstances, is it appropriate, is it realistic for scientists or anyone else to call for a "smaller" Forest Products Industry?

TELEVISION COVERAGE: MORE CONFUSING RHETORIC

It is bad enough when members of the print media march to the drumbeat of antiforestry advocates; it is far worse when television follows in lock step.

As with the print media, some television programmers pursue an agenda which, intentionally or not, fosters confrontation and spreads confusion. The contrived issue of harvesting is given all sorts of attention, while the real issue of growing trees is ignored.

Network Sound-Bites

The following are some highlights of television coverage of forestland issues in recent years. On 26 April 1989, one network introduced the northern spotted owl with the explanation that it was "threatened" and then displayed a banner on the screen that read, "Timber Battle."[13] Viewers were led to believe that the "Timber Battle" issue was about "thousands of acres of virgin forest" and that "the spotted owl lives in the same old-growth trees that feed the Northwest sawmills." Simplistic *and* misinformed!

The morning sound-bite was followed that evening by a story with the introduction: "I want to make some points on how very complicated and immensely frustrating environmental issues can be." The heavy dose of confusing rhetoric included such examples as "Trees are being cut down at an alarming rate."[14] The commentator goes on: "[This] could mean that once again man is going to contribute to the elimination of the ozone layer which will let the sun's

most dangerous rays in. . . . That is where the spotted owl comes in."
Even if there were a connection, that's quite a bit of pressure to put
on a shy little bird.

As the scene of the newscast shifts to the Far West, the harvesting nonissue joins the owl vs. logger one. We see trees being
felled while loggers with chainsaws over their shoulders loom large
and ominous. The commentator then states, "But now environmentalists say it's got to stop or at least slow down." And another
representative of an advocacy group observes, "We're at the wall—
we're running out of timber." Misinformation verging on nonsense!

On the other hand, he may have a point, for should he and his
colleagues have their way and designated reserves continue to bur-
geon, we will be "running out of timber."

MacNeil/Lehrer News Hour

In May 1991, a judge in Seattle ruled in favor of the northern
spotted owl and decreed that millions of acres of renewable forestland ecosystems be set aside to "save" the threatened bird. A
momentous decision.

This event did not escape the attention of the venerable,
award-winning MacNeil/Lehrer News Hour. Eleven minutes were
devoted to it. Unfortunately, as the coverage unfolded, it became
apparent that MacNeil and Lehrer were also victims of the propaganda of antiforestry advocates.

Lehrer introduced the subject with, "Next tonight, another kind of
economic debate. On one side, the logging industry of the Pacific Northwest; on the other, a tiny bird that is threatened with extinction."[15]

He then turned to a Seattle public station moderator, who
added to the confusion: "It's been a year since the United States Fish
and Wildlife Service granted Threatened Species status to the owl
under the Endangered Species Act. That means that habitat for the
estimated four thousand remaining owls in Northwest old-growth
forest needs to be preserved."

The moderator's statement is accompanied on the screen by
daylight views of many "tiny, shy, nocturnal" owls in what appears to

be a second-growth forest. The scene switches to an old-growth forest with no owls in sight. The roar of a chainsaw is heard in the background. In reality, this roar is not taking place in old growth. Instead, as the following scene with the continuing roar shows, the chainsaw is being operated by a logger working in an obviously immature stand of second growth. The fact is, since there just aren't many logging scenes of any sort available today in old-growth forests the camera crew was forced to improvise.

The logging appears to be taking place on private, nonindustrial forestland. Neither the Forest Service nor large private operators would approve harvesting such immature stands.

But it is true that small mill operators in the Far West are being forced to scramble for raw material in the private sector. They have little choice. They must meet interest payments and other fixed costs and make every effort to retain their skilled work crews. It is a throwback to the bad old days when it was necessary to harvest the raw material and move on without regard for stewardship. Of course, many of the smaller operators who have depended on National Forest raw material are simply disappearing.

For their part, the large corporations with fee ownership of forestland are curtailing one mill after another and in some cases abandoning mills altogether in order to protect their young forests.

The great irony of the success of antiforestry advocates is that nothing could be more detrimental to trees and forestlands, or, of course, more economically unsound.

Next on the screen is an environmental activist who states, "You have to remember that 90 percent of the ancient forest has already been logged. We don't call it balance when the timber industry has already eaten 90 percent of the pie and they want to share equally in the last piece." More misinformation.

Industry does not eat the pie. Consumers—all of us, including this antiforestry advocate—eat the pie. Industry is the server. Following this, Oregon congressman Peter DeFazio asserts, "The bottom line is profit and greed." The word greed, greatly overused, in this context describes consumers, the congressman included.

Throughout the eleven minutes, the moderator played the

straight man. The Forest Products Industry, as such, was never mentioned. To link "products"—our essential daily needs—with "timber" is prohibited by antiforestry advocates.

Ancient Forests: Rage Over Trees

The Opal Creek drainage in the Willamette National Forest in Oregon is the location for this one-hour Audubon special, and the title, "Ancient Forests: Rage Over Trees," suggests its theme—confrontation.[16] The opening scene immediately sets up the harvesting "issue" with a shouting match between a representative of the Forest Products Industry and a young mother. No trouble telling who plays the "heavy" here.

Jerry Franklin, the father of "New Forestry," is then introduced by the narrator, Paul Newman: "He has been studying old growth for twenty years and is the foremost U.S. authority on ancient forests." Dr. Franklin speaks to a preservationist as they stroll through an old-growth forest: "There are some people who are totally utilitarian, you know, and if they aren't putting it to use in some very practical human-directed way they aren't happy with it."

Indeed, over 260 million Americans probably feel that way, that is, if they give the matter any thought.

Dr. Franklin continues:

> It is interesting, the rate of timber cutting that's going on here in Oregon and Washington actually exceeds the percentage in terms of rate of cutover that's going on in the tropical regions. And so you know we hear all the time about the rate of deforestation of the tropical forests and it's tremendous and it's a catastrophe but we have a much higher rate of forest cutting going on here in the Pacific Northwest. So all you gotta do is fly over the landscape and you have a real appreciation of how rapidly we're cutting it over. So Brazil has nothing on the Northwest.

Rhetoric can't get much more misleading that this. Comparing industrial forestry in the Far West to clearing land in Brazil is like comparing the proverbial apples and oranges. For the most part, tropical forest clearing is for subsistence farming and grazing, not for raw material. Moreover, in the United States, industrial forestry practices, with few exceptions, call for renewing the forestland after harvest. State regulations in California, Oregon, and Washington leave scant opportunity to do otherwise.

The camera crew then focuses on a group of volunteers lending a hand with trail maintenance in the Opal Creek drainage. The dialogue shifts to a nasty attempt at belittling the Forest Service trail program. One volunteer expert, for example, says of the Willamette National Forest:

> In the next ten years the Willamette plan right now proposes to build 20 miles of trail and something like 1,000 miles of road. . . About 180 miles of trail or something like that will be disrupted by roading in the next ten years. . . We got 20 miles of new trail, we're going to lose 180 miles of existing trail.

Actually trails are being built at an impressive rate, in Oregon as well as elsewhere. The Willamette National Forest plan, completed in 1990, calls for some 560 miles of new trails throughout the seven ranger districts over the next ten years with 90 miles in the Detroit Ranger district, which includes: Opal Creek.[17]

Moreover, since there will be less harvesting on Oregon forestland during this time, planners do not expect to lose any established trails. In fact, during the 1980s, when there was considerably more harvesting than what is expected in the 1990s, there was no net loss in the Oregon forestland trail system, although it was necessary to reconstruct and reroute some trails. Overall, the National Forest trail system as of fiscal year 1992 totals 120,284 miles, an increase of 3,699 miles over 1991. For Oregon the trail system totals 11,008 miles, an increase of 478 miles over 1991.

The old-growth discussion continues and the preservationist says to Dr. Franklin: "The timber industry is very fond of saying we have locked up 2 million acres of old growth. What does 2 million acres of wilderness really give us?" Dr. Franklin replies:

I can tell you for sure, it doesn't give you very much at all in the way of old-growth Douglas-fir/hemlock forests. But you know the wilderness boundaries better than I do and you know perfectly well that they have been drawn to exclude large areas of old-growth timber. And that's truer in Oregon than it is in Washington. In Washington State there are several wilderness areas which do include low elevation timber but that's not true in Oregon. For all of its interest in conservation, Oregon has almost none of its legacy preserved.

True, or more misleading rhetoric? Federal land set-asides are a matter of public record. You will recall that designated reserves include forestlands contained in Wilderness Areas, National Parks, and other designated set-asides, of which there are many.

In addition to the public lands of the National Forest System which have been reserved in Oregon, we have significant federal land set-asides by other agencies, including the Bureau of Land Management, the U.S. Fish and Wildlife Service, and the National Park Service. These contain nonforested as well as forested land.

In Oregon, the most significant public land set-asides alone total 6.654 million acres. This compares with 4.868 million acres in the state of Washington. The total area of federal set-asides for all fifty states is slightly under a quarter billion acres (247.492)—nearly the total land area of the fifteen states in the Northeast quadrant (256.988).

To compare in round numbers federal set-asides in Oregon and Washington: In the National Forest System, Oregon has 2.649 million acres, Washington has 2.745 million. As indicated earlier, this is largely fully stocked forestland, heavy to old growth, which is now part of what we call designated reserves and therefore removed from the resource data base.

Within the Bureau of Land Management (BLM), Oregon has 3.240 million acres of set-asides as opposed to only .005 in Washington. These BLM lands include 2.4 million acres of Oregon & California railroad grant lands in western Oregon, most of which is fully

stocked forestland. (These O&C lands were revested to federal ownership when the railroad was not constructed.)

Comparing set-asides in the Fish & Wildlife Service, Oregon
has .567 million acres and Washington .173; in the National Park
Service, Oregon has very little acreage compared to Washington—
.198 million vs. 1.945.

For readers interested in more detail, refer to Table 8 in the
Appendix. Also included in Table 9 in the Appendix is a list of all
states showing acreage for each, along with acreage owned by the
federal government.

What does all this suggest? At the rate Congress and the courts
are expanding the designated reserves, will there be anything left to
meet the growing needs of this country and the world? As of now,
we can meet our noncommercial needs with little difficulty. The
ability to nurture renewable forestland ecosystems, however, has
been taken out of the hands of professional foresters.

For example, Oregon State University reports that with the
Forest Service and BLM land-use allocations in effect in western
Oregon (1990), fewer than one out of three acres are available for
intensive forestland management.[18] Their findings indicate that
only 29 percent of forestland is available for intensive management
and only another 10 percent is available for management with "view-
shed" restrictions.

In Washington State the abuse of the National Forest System
charter is much more drastic. For example, on the 1.7 million acre
Mount Baker-Snoqualmie National Forest, only thirty thousand
acres, or less than 2 percent, is available for raw material. In the
decade of the 1980s, the annual allowable sales quantity goal on this
forest dropped from 290 million board feet to twenty.[19]

Returning to the program we see another well-meaning scientist, Chris Maser, listening to the host preservationist, who says,
"The Forest Service says that these old-growth forests are full of rotten trees and rotten wood and it's a waste to let them sit here and not
cut them down and take it out." Maser replies, "There's nothing
wasted out here."

That of course is true, if we assume that Maser, a trained zoologist and aspiring ecologist, means nothing will be wasted if we defer to nature and allow insects, disease, windthrow, and wildfire to harvest our renewable forestland ecosystems.

Next, standing before a background of a recently replanted area, Maser and the host preservationist share their views. Maser observes, "This is a tree farm; we call it reforestation." And he proceeds to denigrate the remarkable advances in forestland management by professional foresters. "There are no snags," Maser states, "there are no large logs, and when we lose these, we lose a lot of the habitat structure. The cavity nesting birds are not going to be represented here. The small mammals that take the mycorrhiza spores and disperse them throughout the forest are not going to be represented here." He continues, "A forest is not trees. . . trees do not make a forest." Interesting. What difference then does it make whether or not you harvest?

In sponsoring this sort of misleading commentary, the National Audubon Society, a preeminent conservation organization, has descended to using shoddy tactics.

Nova, "Return to Mt. St. Helens"

This one-hour program marking the ten-year anniversary of the eruption is almost consistent with the excellence we have come to expect from Nova.[20] It unravels many of the misperceptions and misleading rhetoric on the subject of forestland ecosystems.

The viewer is shown spectacular footage within the restricted zone inside the crater and throughout the surrounding area of the National Volcanic Monument. Scientists share their views, their theories, and their findings in layman's terms in fascinating detail.

A good portion of the program is devoted to forestland conservation and management and to explaining how both plant life and wildlife are adapting to the entirely new ecosystem. Some of the most interesting coverage deals with studies of insects, animals, and fish.

The first perception these scientists debunk is "fragile ecosystem." Any viewer of this Nova program would be impressed with how inappropriate that concept is in the real world. The same can be said for the idea that harvesting does away with "biodiversity" which, according to this blue-ribbon team of leading scientists, is immune even to volcanic eruptions.

The argument of antiforestry advocates that harvesting a forest "destroys" plant and wildlife habitat, including fisheries, is also exposed in this program, which attests to the resilient and dynamic characteristics of our forestlands.

The environmental activist on the MacNeil/Lehrer News Hour, referring to the spotted owls, remarked, "You have to save habitat to save species." Once again, this Nova program proves otherwise; throughout the program we see over a thousand species, from tiny insects to herds of elk, thriving in what the scientists initially presumed to be a sterile habitat.

Unlike the inflammatory rhetoric of the antiforestry advocates which flies in the face of biological science, Nova offers supportive data for scientific findings. We see it in the field and in the laboratory. One scientist, having worked on insect fall-out on other mountains, expressed his surprise at the extent of diversity within the blast zone. The scene then switches to mice and burrowing gophers in this most hostile appearing ecosystem. It's eerie, as if astronauts were finding animal life on the moon.

But the most impressive footage was of a healthy looking herd of elk. These beautiful animals, which some insist can be found only in old growth, are shown gamboling through what looks like a Weyerhaeuser Tree Farm.

The startling Nova coverage doesn't end here. The scientists track the elk from the tree farm out into barren ash plains while we are told how their body systems deal with such a widely fluctuating environment. At this point, they are many miles from old-growth forestlands but relatively close to their browse.

As for fisheries, we witness various methods of documentation, including scientists donning wet suits and taking fish counts in streams within the blast zone. The narrator details their findings: "Far from being damaged by the invasion of half a forest into their stream, the fish population flourished. Exposed to open sky and hot summer sun, they have faced water temperatures far higher than ecologists thought they could tolerate. The fish keep cool by hiding in pools of water cooled by the shadow of logs."

When logs and other debris are in the streams, the fish habitat is actually enhanced. Science has set the record straight about multiple-use; in working commercial forestlands, as with volcanic eruptions, when logs and debris end up in stream beds, fish habitats are not "destroyed."

Once again we are introduced to the father of New Forestry, Dr. Franklin, who here admits that his "ideas and concepts and theories" don't hold up well in real life. As footage on the screen reveals the destruction within the blast zone, Dr. Franklin says:

> We imagined that this was going to be a very sterile place where everything had been destroyed and everything was dead. And what we found instead was that a lot of life, a lot of organisms, a lot of biological structures were left behind here. So what we had here was an incredible richness of things to look at. And every time we turned around there was something new and exciting and different and surprising. We were being blind-sided all the time. And I described it as being like a kid in a candy store because there were so many things going on here. And I always loved the way in which we'd develop our ideas and concepts and theories and then would come in here and nature would just knock 'em on their pins.

Why can Dr. Franklin not experience the same "excitement" in renewal after harvest? And why does he overlook the science and technology of industrial and nonindustrial forestry so painstakingly developed over the past fifty years?

On the one hand, he is overwhelmed by how nature springs back after a volcanic blast, while, on the other, he maintains that renewable forestland ecosystems are unable to recover from a harvest—unless his theory of New Forestry is applied.

Questions have been raised by professional foresters about whether on-the-ground research is adequate to support the adoption of the New

Forestry concept on a commercial scale. The Nova findings certainly cast doubt on both the concept and the need for New Forestry.

Rather, our renewable resources exhibit their natural resilience and dynamic adaptability to change in a most impressive manner, and when helped along by ecological science and technology, the renewal process is even more impressive. Unfortunately, the Nova program wrap-up lapses into a commercial for New Forestry under the guise of "biological legacy." The narrator, in summary, says:

> Out of all the diverse studies of nature's recovery at Mount St. Helens, one unifying concept has emerged: biological legacy. Scientists now understand the importance of what survives and what remains. They know the profound value of biological oases protected by ice and snow, of spiders and seeds that blow in on the wind, of sunlight and decay, of animals that turn over the soil, and nutrients gleaned from the dead that speed the recovery of life. That is biological legacy.

Then he notes that this legacy has compelled the father of New Forestry to "take a different view of nature." And the father of New Forestry restates his amazement:

> The magnitude of this disturbance and the importance of legacies here really rubbed our nose in it. And then after we'd spent a couple years here recognizing this, we sort of went back to other kinds of disturbances and took another look at them.

Not, unfortunately, a revisit to reality but another serving of fantasy:

> The commercial forests around St. Helens regularly sustain such disturbances. Extinguished and recreated by clearcut logging where every usable fragment of timber is scoured from the ground. Logging is the main industry of the Northwest. After a century of it, only 5 or 10 percent of America's untouched old-growth

forests remain. What grows instead is man-made. A uniform crop of little diversity. Tended for fifty years, then cut down again. Old-growth forests are trees hundreds of years old. Rich with beauty and ecological value but that value competes with timber values. Preserve them or consume them.

As the Nova program winds down, between the narrator and Dr. Franklin, we slip more and more into a forestland fantasy. Both do a first-rate job of demeaning forestland management by professional foresters. Dr. Franklin avers:

> We don't have to make such an incredible choice between either, you know—very simplistic agricultural type forestry systems on the one hand, where we cut everything and haul it off or, on the other hand, strictly preserve it, just lock it all up and throw away the key. What the legacy idea does is begin to give us some idea of how to practice forestry. So we can have commodities in reasonable amounts and at the same time we can retain these ecological values or at least a lot of them within our managed landscape.

I have yet to meet a professional forester who would agree that his or her work was "simplistic." As for the "legacy idea" which will "give us some idea of how to practice forestry," a professional forester would be able to make such a statement only if he or she were emerging from Sleepy Hollow after a fifty-year nap.

And why does the narrator demean professional forestry practices—"every useable fragment of timber is scoured from the ground"—for utilizing all forestland resources? Does he wish to revert to the wasteful practices of the nineteenth century?

Sadly, the program, which had been so instructive, is moving from facts into advocacy. True, logging is a vital function of the Forest Products Industry; but it is only one of many steps necessary to satisfy the needs of society. And to describe logging as "the main industry of the Northwest" is simply foolish.

The narrator seems totally unaware of forestland renewal, of the myriad vigorous stands of lush new forests. And he ignores as well the millions of acres of old growth "rich with beauty and ecological value" which are set aside and protected by law in designated reserves.

Why these unending claims that harvesting destroys forestland ecosystems when we know it is not possible to destroy ecosystems, as witness their survival even when a volcanic blast transforms vast areas of forestland into virtual moonscapes?

When our best scientists find extraordinary biological diversity in the barren blast zone of Mount St. Helens, you would naturally expect them to find even more in a renewed forestland ecosystem, yet the narrator speaks of a "uniform crop of little diversity."

On the subject of diversity, original forestlands on the west side of the Cascade Range contain monotonous concentrations of Douglas-fir. Renewed forestlands contain stands heavy to Douglas-fir precisely because professional foresters are following nature's example. They take full advantage of the fact that this valuable species is favored by the climate and soil of the region.

Why then should stands heavy to one species be denigrated in the Nova program? Visitors never complain about the monoculture of redwood forestlands as they drive by only redwoods for more than thirty miles along the Avenue of the Giants. Must we be bound by what antiforestry advocates and well-meaning scientists dictate as the "correct" way to care for our trees and forestlands?

REALITY VS. PERCEPTION

In the introduction to this chapter, science-based enabling legislation was compared to the more recent gridlock, restrictive legislation. There is room for common ground and we can solve regulatory gridlock, but we must recognize that much of what is popping out of Pandora's Box has to do with perceptions as opposed to reality.

To find common ground, we must look at the ecological evidence and agree that sensible stewardship is not in question. To repeat, the renewable resources that make up our environment have inherent characteristics; they are resilient, dynamic, changeable, and changing. This renewal process for air, water, soil, flora, and fauna has been on-going for eons, and over the last fifty years has received a tremendous boost from huge investments in science, technology, and skills dealing with forestland renewal.

Antiforestry advocates have taken a fork in the road which most professional resource managers view as a wrong turn. Common ground will not likely be found at the end of this road; rather, it appears that vested interests in politics and fund raising hold greater sway than nurturing our renewable resources.

When antiforestry advocates paper over ecological evidence with the perception that our renewable resources are essentially non-renewable, that they are subject to permanent degradation and decline and always damaged, if not destroyed, by professional management, permanent gridlock is the result.

To break the prevailing gridlock we must turn to one of our nation's greatest strengths—the unfailing ability of ordinary citizens to rise to the occasion and do the right thing. In the case of nurturing our renewable resources, it is to recognize what *is* and what *is not* stewardship.

3

FORESTLAND ECOSYSTEMS:
STEWARDSHIP

ECOSYSTEM MANAGEMENT TODAY calls for a kinder, gentler concept of forestland conservation and management than yesteryear, but we should not belittle the profound dedication of the early conservation pioneers nor lose sight of the fundamental precepts laid down by them. These must be kept in mind as we review what *is* and what *is not* stewardship.

A COMMON VISION LEADS TO A SUPPLY MIRACLE

When foresters and forestlands were brought together in 1905 by the Transfer Act, professional forestry, then in its infancy, gave welcome to citizen conservationists. A common vision developed, first, to protect forestlands and watersheds from wildfire, and second, to overcome the practice of cut-and-run. This early vision was integral to the emerging conservation ethic and laid the groundwork for the outstanding achievements involving cooperative forestry, multiple-use, and high-tech industrial forestry. It also made possible a supply miracle. None of this was accomplished overnight.

For three-quarters of a century, from 1875 to 1950, citizens were dismayed at the exploitive logging practices. Cries of "timber famine" actually began as far back as the eighteenth century, and they continued throughout the nineteenth, and were still being heard after World War II. Since then, however, the supply miracle has stilled these cries without in any way depriving consumers.

The modern supply miracle came about through huge investments in research and state-of-the-art technology, which so improved our forestlands and converting plants that they became the standard of comparison worldwide.

But the most amazing aspect is that it occurred in just fifty years, despite restrictive legislation. No less impressive is that some 67 percent of our original colonial forestlands are still intact. This in itself is proof of the inherent characteristics of our renewable resources. They are indeed resilient, dynamic, and, in recent years, changeable for the better when nurtured.

Nevertheless, regulatory gridlock, based upon a simplistic misunderstanding of the nature of our renewable resources, has taken its toll. Citizen conservationists are often at odds with the high-tech of industrial forestry. Moreover, instead of professionals, members of Congress and judges determine how forestland conservation and management should proceed. It is much like the bad old days prior to 1905 when professional foresters were separated from the forestlands.

COLONEL WILLIAM B. GREELEY: A MAN OF VISION

Colonel Greeley was a man of extraordinary vision. He had a knack for dealing with reality and the fundamentals of forestland stewardship. Greeley retired from the West Coast Lumbermens Association in 1945, but continued to fight for another ten years to turn the Forest Products Industry away from its old destructive past to a new era of renewing forestland ecosystems.

Given his record of over fifty years as a professional forester, Greeley's impact was of singular importance. He was a major influence in launching and implementing not only cooperative forestry but also high-tech industrial forestry. And his vision of a supply miracle was entirely accurate. At a 1945 dinner in his honor in Portland, Oregon, he said:

In the next ten to fifteen years, I can very clearly see taking shape, new ideas, new industrial processes, new uses of wood. We are going to see a great alchemy of forestry, of utilization of new products based upon the marvelous growing power of the Douglas-fir forest. I expect to live to see West Coast forestry become the outstanding example of forest economy in all these aspects.[1]

While Greeley maintained that the preservation ethic was an important part of forestland stewardship, he would undoubtedly have taken a dim view of the sort of restrictive legislation that has encouraged a "let-it-burn" policy to raze millions of acres of forestlands in the public sector, and allowed wildfires to rampage over private holdings as well. His experience with wildfires in the Far West made an indelible impression on him.

WILDFIRES ARE KILLERS

Wildfires, as opposed to controlled fires, are killers! In the most brutal fashion they end the American dream. Human lives, homes, businesses, and wildlife are consumed.

In the first chapter of Greeley's book, *Forests and Men*, he relates his bitter experience with wildfires. In 1908, at age twenty-nine, Greeley was put in charge of the newly created 25 million acre District 1 (now Region 1) of the Forest Service in Idaho and Montana. Two summers later, wildfires burned 3 million acres of his district. Seventy-eight fire fighters and seven citizens died. Greeley recalls, "To a young forester, thrown by chance into a critically responsible spot on a hot front, that summer of 1910 brought home the hard realities of our job. Ideals, glorified by 'the golden haze of student days' and the enthusiasm of inspired leaders, suddenly came down to earth."[2]

An estimated 8 billion board feet of old-growth forests were burned, including some of the finest stands of Idaho white pine in the Panhandle. Thus, a single wildfire loss was nearly equal to the total harvest on our National Forest lands in fiscal year 1991. Greeley went on to say:

And I had to face the bitter lessons of defeat. For the first time I understood in cold terms the size of the job cut out for us. We had to overcome the habits and practices of a people who had taken their forests for granted—for two and a half centuries. And we had to engineer and organize the best resistance man can devise against terrific natural forces which at times are overpowering. From that time forward, "smoke in the woods" has been my yardstick of progress in American forestry.[3]

Eighty years after the 1910 wildfires in Idaho and Montana similar circumstances pertain. Northwestern Montana today is replete with mile after mile of overmature forest stands; in every direction, there are endless dying, dead, and down lodgepole pine, and tens of thousands of acres of beetle kill.

We are familiar with the devastation wrought by the 1988 wildfires in Yellowstone National Park: thirty thousand fire fighters were on the job, ten were killed. Many observers agreed that the park's unmanaged forestland ecosystem, where nature was allowed to take its course, was an accident waiting to happen.

It is no surprise that in recent years wildfires have raged through the same forests that Greeley tried unsuccessfully to save in 1910. We will undoubtedly suffer much more of the same in the near future.

For example, throughout the area burned in 1910, salvage sales are critically needed to clean up distressed and overmature forests. There is no other way to renew these forestlands so that watersheds and wildlife can be protected and a place of recreation maintained. It is a fundamental decision. Are we going to manage these forestlands and give their resilient nature a chance or are we going to allow wildfire, insects, and disease to take their toll?

The situation in Idaho and Montana is already serious with 7.8 million acres in Idaho set aside in various reserves and 6.4 million in Montana. In Idaho, this includes 4 million acres of Wilderness within National Forests; in Montana, 3.4 million acres. In these vast Wilderness areas, management is precluded by law.

Meanwhile, members of Congress from both states and antiforestry advocacy groups are aggressively pushing bills which will significantly expand designated reserves, add another layer of grid-lock, and expand a vicious precedent for negative stewardship. On 26 March 1992, the Senate passed the Montana Wilderness bill (S. 1696, introduced by Senator Max Baucus) by a vote of 75 to 22. This bill designates twenty-seven new Wilderness Areas, totaling some 1.2 million acres. The bill also creates 285,000 acres of special management areas and sets aside another 715,000 acres for "future congressional action."

This kind of legislated stewardship is having an unbelievably negative impact on reasonable management efforts. But it is pack-aged in such a way that it is viewed, superficially at least, as the right thing to do by the average citizen.

The situation in Idaho and Montana is no isolated instance. Indeed, our conservation ethic which honestly seeks to take care of our forestlands is under attack throughout the country. The Blue Mountains of northeastern Oregon offer a similar case history:

> An April 1991 Forest Service "Blue Mountains Forest Health Report" showed that 53 percent of the forest-land—nearly 2 million acres—in the Malheur, Umatil-la, and Wallowa-Whitman national forests was dead or defoliated by insects in 1990. . . . Scientists now believe that about 4 million acres in the Blue Mountains are dead or dying—almost two-thirds of all federal, state, and private forestland in the region.[4]

An industrial forester based there makes this observation: "Whether your values are aesthetic, spiritual, or economic, these forests are not providing it."

A Forest Service scientist and director, Tom Quigley, of the Blue Mountain Resource Institute said, "It took us one hundred years to get into this situation, and it's probably going to take us as long to get out of this." Interesting but probably not accurate. Inevitably, wildfires will clean up the mess in a matter of days. But at what cost in human life and wildlife habitat?

Wildfires and true conservation of trees and forestlands are on opposite teams. But the perception persists in the minds of the general public that controlling wildfire is a primary reason for the widely publicized forestland health emergency.

This makes little sense. In managed forests in the private sector in the United States, as well as those cited earlier in Germany, catastrophic fires are rare because fuel loads are not allowed to accumulate. When wildfires do occur in the private sector, it is usually as a result of uncontrollable wildfires in the public sector. This is because the unmanaged fuel loads are in the public sector, not in managed forestlands in the private sector.

One of the reasons the issue of the forestland health emergency and wildfires is confusing is because the difference between destructive wildfires and beneficial controlled fires has been inadequately explained. For many years, controlled fires have been used to eliminate the buildup of fuel loads. But these controlled fires, or any other management technique, are not allowed in designated reserves, and too often not in the public sector, because of lack of resources and because of the distaste for them by the general public.

SNAGS ARE KILLERS

Snags are those dead trees still standing, usually as bleak testimony of some previous disaster. When they rot, they sometimes provide homes for cavity nesters—birds and small animals. Snags, like wildfires, are dangerous and all too often, killers.

There are far better means of providing for cavity nesting birds than the current regulations that require the protection of snags. As an example, in that Bavarian Sauerlach forest just outside Munich, there are no snags; instead, there are some 1,500 birdhouses built by foresters and volunteers. (Note photo 1.)

Is this farfetched? Not according to wildlife experts in South Carolina's Francis Marion National Forest who have already discovered that fabricated nesting structures replacing those destroyed by Hurricane Hugo in 1989 are preferred by the endangered red cockaded woodpecker.

Whenever there is a major electrical storm, particularly in the Far West, snags are nature's made-to-order lightning rods. And that is just one of their dangers. Loggers call them "widowmakers." Snags often drop broken limbs and debris into the crowns of live trees, and when felling is underway, workers can be killed by these death-dealing missiles from on high.

Widowmakers, like wildfire, are indiscriminate killers. Recreationists are also in peril in old-growth forest stands or harvested areas where snags have been left. Thus, in the fight to drive smoke from our forestlands, it became standard operating procedure to fell all snags—that is, until laws were passed protecting them.

The most graphic demonstration of the wisdom of removing snags came from the Tillamook, Oregon, fire of 1933. This wildfire showered ash on metropolitan Portland and took twelve lives, making headlines across the country. The first burn scorched some 311,000 acres and was a classic "fire hurricane." In one day, 270,000 acres of prime, old-growth Douglas-fir were consumed.

Then a wildfire pattern designed by the devil himself took over. In the burned area, the combination of snags, tinder-dry fuel, and lightning sparked another wildfire in the Tillamook region in 1939, and again in 1945, and again in 1951.

Professional foresters and concerned citizens were finally galvanized into action by this destructive cycle. In 1948, Oregonians passed a $10 million Oregon Forest Rehabilitation and Restoration Bond Issue. Reforestation of the Tillamook began immediately and was completed in 1973.

Progress in the first two years was so impressive that William Greeley remarked, "The restoration of the Tillamook will rank. . . as one of the great human achievements in engineered conservation."[5] Among several observations, he said natural reseeding and early hand-planting efforts were unsuccessful because young growth was wiped out by the next wildfire. He also noted that "no seeding or planting by the hand of man could be done with any assurance of survival until the unbelievable fire hazard created by millions of standing dead trees had been conquered."[6]

Access roads and "snag free corridors" 200 to 2,000 feet wide were made possible by the bond issue, which also financed salvage logging, wildfire detection, and control. Today the Tillamook forest is a beautiful, lush green testimonial to what happens when concerned citizens and professionals join hands to nurture forestlands.

But with the advent of restrictive legislation, snags have become a part of forestland stewardship; common sense is ignored. The unholy marriage of snags and lightning strikes cannot be overly stressed. Every year lightning starts some ten thousand wildfires in our forestlands. The price for a bird nest in a snag is too often a human life or a catastrophic wildfire that consumes all living things.

Wildfires also kill countless large animals, including grizzly bears. Grizzlies are an important part of our heritage and we should and can manage bear habitat along with renewable forestland ecosystems. But U.S. Fish and Wildlife Service personnel are pushing those grizzlies right into the backyards of rural families in the Northern Rockies.

GRIZZLY BEARS. . . KILLERS?

In 1988 the management of grizzly habitat became actually frightening for families like the Vincents of Libby, Montana.[7] The home of Bruce and Patti Jo Vincent and their four children is nestled within the Cabinet-Yaak ecosystem. The Vincents, along with three of Bruce's brothers, have for many years managed a family logging business.

Thanks to a local newspaper, Bruce became aware of U.S. Fish and Wildlife Service plans to "augment" the grizzly population in their rural area. Not knowing exactly what that entailed, Bruce and Patti Jo attended a local meeting where they learned that augment meant restoring the grizzly population to its historic level.

In June 1991, Bruce Vincent related this news to fellow loggers at a Southeastern Wood Producers Association meeting in Waycross, Georgia. The Fish & Wildlife representatives had indicated they were only "holding public hearings as a matter of courtesy."[8] Apparently, the opinion of local residents was not considered important.

Angered, these locals "joined hands and said *no*. Not *no* to the grizzly bear, but *no* to the management plan. We were able to stop the funding for the plan for one year. Then we started a community involvement committee that works for a common future for both us and the bear."[9]

Bruce and Patti Jo took small comfort from the federal representatives at the initial meeting. When Bruce inquired, "What is the historic level of the grizzly population in our rural area?" The answer was, "We don't know. . . .At least we have to establish a biogenetic pool." When Bruce asked, "How many is that?" Again, the answer was, "We don't know, but we think ninety to 120 of them in your area will do it."[10]

Bruce then asked, "How many of them are there now?" Again, the answer was, "We don't know, but we think four." Bruce said he had grown up in the area and thought that ninety to 120 bears might present a problem. He was told not to worry, the Vincent family would only have to adjust their way of life a "little bit." But since their home was within the "human-grizzly conflict zone," they "would have to tie bells on their shoes to warn the bears away."

When Patti Jo asked if she would have to tie bells on the shoes of the Vincent children when they went out to play, the reply was, "Yes, and that should keep the bears away." If "we find a bad bear, we'll deal with it." At that point, the local parents learned the difference between a bad bear and a good bear—"Bad bears have bells in their poop."

To be fair, although grizzly bears have been known to drag young women out of their sleeping bags in Glacier National Park and kill them, this doesn't mean that all grizzlies are killers. Grizzlies, like wolves, have been given a bad reputation in literature. Rural citizens like the Vincents make it clear that they have "no problem with the bears themselves because that's part of living there. But the recovery program didn't take us into account, and we have some problem with that." The Vincents were trying to explain something about reality, but anyone who attempts to part the curtain of perceptions is labeled a troublemaker.

And loggers? What do they know? For one thing, they know the difference between reality and fantasy. They know they can contribute to stewardship if given half a chance. For example, to ease the desperately high unemployment problem in their communities, these loggers could go back to work cleaning up the forestland mess in their backyard. But that is against the law, because grizzly bear habitat does not allow management of our renewable forestland ecosystems.

Today, about 25 percent of the 3 million acres burned in 1910 have been turned into a tinder box by insects and disease in an over-mature forest that is highly vulnerable to lightning strikes and wild-fires. And we can only wait for a recurrence of the catastrophic wild-fires that destroy bear habitat as well wreaking havoc on humans.

STEWARDSHIP AND ECONOMICS

Antiforestry advocates would have us believe that professional forestry is only concerned with profits. Forest Products Industry operators are naturally concerned with profits and commerical bene-fits. It can't be otherwise. Profits, after all, determine whether oper-ators survive or become part of the high rate of industry attrition. Eliminating waste, increasing productivity, and adding as much commmercial value as possible are critical to survival.

The perception of stewardship economics which holds that com-mercial values rule out noncommercial values is simplistic. In reality, commerical and noncommercial benefits are not mutually exclusive. They can and should be mutually beneficial. Unlike the hard-pressed Forest Products Industry, the environmental industry is one of the fastest growing industries in the world. In 1875 there was one citizens conservation organization. In the 1994 Conservation Directory pub-lished by the National Wildlife Federation, there were 674 listings under the section "International, National and Regional Organiza-tions." They are blooming like flowers in spring.

Of course, there will be attrition in the environmental industry; attrition, however, is something the Group of 10 has never had to face. By and large, over the past thirty years, each member of the Group of 10, the leading environmental organizations (listed in Table 14 in the Appendix), has enjoyed growing membership lists, expanding budgets, and close cooperation with Congress, i.e., power.

Stewardship Economics and Attrition

And the Forest Products Industry? How have the organizations who worked the supply miracle been doing? Though most of the leaders survive, the attrition rate is high and members of the top ten continue to fall by the wayside.

From the 1965 *Fortune 500* listing, three of the ten leaders in the Forest Products Industry have gone: #3 Crown Zellerbach, #4 St. Regis, and #9 U.S. Plywood. From the 1970 listing, #9 Evans Products and #10 Diamond International are gone.

Other major organizations which have disappeared or been absorbed include Georgia Kraft, Continental Can, Brunswick Pulp & Paper, Hoerner Waldorf, Hammermill, American Forest Products, Great Northern Nekoosa, Masonite, Arcata National, and Southwest Forest. Many more, of course, were never listed on the *Fortune 500*.

Leading companies in the Forest Products Industry are noticed when they fail, but these cases are far outnumbered by tens of thousands of operations that have also floundered over the past fifty years and been given scant notice. These smaller, often family-owned operations were also doing their best to satisfy consumer demand, but were unable to survive the fierce competition and regulatory restrictions.

Stewardship Economics and Fund Raising

Attrition, however, is rare in the environmental industry. Money pours into the coffers of the various advocacy groups from massive fund-raising campaigns including frequent direct mail and other devices, and no risk-taking is involved in their activities. Evidently, too, it is not considered reprehensible to use confusing rhetoric to gain the attention of concerned citizens: "They are cutting the last of the old growth . . . they are destroying critical habitat. . . they are being subsidized by taxpayers," and on and on goes the plaint.

The Group of 10 is tightly knit. Leaders meet regularly and stay in close touch. They share mailing lists, and when individuals send contributions to one, they will likely receive fund-raising requests from the others.

Forestland Stewardship and Greed

The word "greed" is often used in these mailings to describe Forest Products Industry operators. Yet no one can be so aggressive as the Group of 10 when it comes to chasing the dollar. By the time a recipient reads through the single-spaced, multipage, fund-raising letters, all of the emotional buttons have been pushed and money has been asked for in various ways many times over.

Leaders of these advocacy groups lose no opportunity to raise money. Immediately after the Exxon Valdez oil spill in Prince William Sound, a feeding frenzy developed. Millions of solicitations poured out.

Not long after came the news from the *Wall Street Journal*: "The world's biggest oil spill—6 million barrels—has largely disappeared from headlines but not from Saudi Arabia's shores."[12] And a follow-up story on 15 October 1991, "A dozen workers struggle with rakes and shovels to rid a beach. . . of mounds of sea grass saturated with acrid crude oil" in Adaffi Bay, Saudi Arabia.[13] The article continued:

> Nearby, a bulldozer scrapes through a foot-thick frost-ing of oil that runs in a wide ribbon as far as the eye can see. In a few days, these efforts will have given a two-mile stretch of this shore, near the port town of Juball, back to the bathers and the birds. That's two miles down—and about 398 miles to go.

This cleanup has been "carried out by a handful of Western contractors working for the U.N.'s International Maritime Organization. Their $6 million budget is drawn from foreign contributions, says Dave Usher, an IMO official. The far smaller 1989 Exxon Valdez spill in Alaska attracted 11,000 workers and a $2.5 billion effort." The ratio between the two budgets is 415 to one.

Now, let us make one thing clear: There *is* an urgent need for environmental activism. All citizens should favor reasonable, balanced, and fair environmental agendas. And those responsible for damaging our environment ought to be brought up short.

The environmental industry, unlike the Forest Products Industry, attracts our young people's attention and interest. Their enthusiasm for environmental protection deserves our support. They are bright, conscientious, dedicated, and energetic—but they are in danger of being misled. The enthusiasm and energy of our young people ought not to be misdirected.

Citizens must be aware of the brand of forestland stewardship these advocacy groups are espousing. Placing our renewable forestland ecosystems in designated reserves is not stewardship. Promoting restrictive legislation that frustrates the charter of the Forest Service—"to furnish a continuous supply of timber for the use and necessities of citizens of the United States"—is not stewardship.

Antiforestry Stewardship

The proposed Clearcutting Restraint Act of 1989 (Rep. John Bryant, D, TX) is an example of antiforestry stewardship. Here is some of the language: "Clearcutting is the practice of completely destroying a forest. . . . Clearcutting is a wasteful and irresponsible practice. . . . This practice must be stopped because it is enormously wasteful. . . .Clearcutting totally eliminates species diversity of plants and animals."

There are nineteen paragraphs in the congressman's two-page introductory statement. The language of the introduction and the language of a fund raiser are interchangeable. Moreover, he states, "I welcome the support today of the Audubon Society, the Wilderness Society, the Environmental Policy Institute and the Friends of the Earth for this legislative initiative." At least he is candid about the source. In 1994, Congressman Bryant introduced the Forest Biodiversity and Clearcutting Prohibition Act (HR 1164) with ninety-four cosponsors.

When the economics of forestland conservation and management are totally ignored, stewardship is not possible. And yet, ignoring such economics is common among antiforestry advocates. In May 1989, I interviewed a representative of the Sierra Club Legal Defense Fund in San Francisco. I was interested in learning more about environmental activist stewardship and its impact on the economy in the Far West. I asked: "Do you hear from representatives of the Forest Products Industry in the Northwest?"

"Yes," the representative replied. "I have had a number of letters and phone calls from industry workers, wives, and even children."

"Well, how do you deal with their concerns?"

"We are not concerned with the economic impact of our work," she said.

Stewardship: The Media and Fund-Raising Ploys

Fund-raising ploys often make no contribution at all to forestland stewardship. A specific example in the *New York Times* gives adequate testimony of this. Some thirty years ago, an architect friend who had worked full-time in the media in New York City informed me, only half in jest, that the motto of this paper was, "All the news that's fit to slant." This is helpful to remember when reading copy, and advertisements about the Forest Service, Forest Products Industry, and industrial forestry. For example, on 6 February 1990, a full-page story appeared as paid advertising which flayed Idaho Senator James A. McClure for introducing a forestland conservation bill dealing with reality.

The headline read, "Sneak Attack on the American Wilderness." A large photo of a beautiful alpine lake was the centerpiece of the advertisement. The caption identified it as "Buck Lake and the Lemhi Mountains of Idaho, part of the Salmon National Forest. . . the Buck Lake region might eventually resemble the clearcut area in the photo below. You can help stop that. Please use the coupons." (*send money!*)[14]

The smaller photo, captioned "The fate of our forests," was allegedly a "redwood clearcut in northern California." As starters, comparing Idaho forestland with redwood forestland just adds to the confusion. Moreover, it would take some sort of madness to attack American Wilderness, since it is protected by federal law. And then, even to imply that forestland around Buck Lake is threatened is dishonest. Buck Lake is in a Restricted Roadless Area, elevation 8,500 feet. This is close to the elevation in this region above which trees cannot survive, so obviously there is nothing of commercial value here. Professional foresters would give this area over to Wilderness without question.

Had this advertisement been published in 1875 when exploitive logging was the rule, some of the information might have been appropriate. Today it's difficult to see how Sierra Club President Richard Cellarius would permit his name to be used to endorse the story. And are Sierra Club members aware of the dishonest assertions being printed in their name? If it were product advertising in the real world, lawsuits would be tripping over each other.

Donna and I visited Buck Lake in August 1992 and duplicated the photo appearing in the Sierra Club advertisement. Buck Lake is in a basin surrounded by peaks over 10,000 feet in elevation (see photo 11 which is looking southeast).

To show the reader what is meant by "treeline" above which very little grows, note photo 12, looking due east, which indicates a little better how vegetation tapers off just above the shoreline of Buck Lake.

If more citizens could see for themselves how forestland conservation and management is working at Buck Lake in Idaho, they might wonder why anyone would ever run such erroneous copy.

But obviously the general public is taken in by these fund-raising efforts, as the ongoing full-page advertisements that appear in the *New York Times* attest. On 24 March 1994, there was another such, with even more inflammatory copy than in the Buck Lake story. The purpose was splashed across the top of the page—"Campaign to End Industrial Forestry." Photos of clearcuts were prominently displayed. And "ecoforestry," as opposed to industrial forestry, was touted along with a five-point plan:

1. An immediate end to all clearcutting
2. An end to all logging and road-building in ancient forests and roadless areas
3. An end to industrial forest management and the "tree farm" practice
4. Substitution of conservation biology for "industrial forestry"
5. Respect for the intrinsic value, integrity, and stability of forest communities.

It seems abundantly clear that the sponsors of this message have little interest in finding common ground. They are the Sierra Club, Rain Forest Action Network, Canada's Future Forest Alliance, and the Ecoforestry Institute.

Sierra Club: Stewardship Then and Now

At its start, the Sierra Club was an impressive conservation organization. Since World War II, however, it has forsaken the vision of John Muir, the man so important in its founding and its first president. His leadership inspired all who shared his preservation ethic, needed then and now.

John Muir believed that the beauty of the high country should be shared. Some directors, however, did not agree when it came to club activities. They thought outings comprising "large parties, necessarily including strangers, were undesirable."[15]

At the 1895 annual Sierra Club meeting, Muir gave his views in an impromptu speech: "Few are altogether deaf to the preaching of pine trees. Their sermons on the mountains go to our hearts; and if people in general could be got into the woods, even for once, to hear the trees speak for themselves, all difficulties in the way of forest preservation would vanish."[16]

His thoughts were at the core of the Sierra Club's articles of incorporation, drawn up in 1892 by University of California Professor J. Henry Sanger and Warren Olney, a San Francisco attorney.

This document, a statement of the basic purposes of the Sierra Club, was signed by twenty-seven recreationists and conservationists on 4 June of that year: "To explore, enjoy and render accessible the mountain regions of the Pacific Coast; to publish authentic information concerning them; to enlist the support and co-operation of the people and government in preserving the forests and other natural features of the Sierra Nevada."[17]

John Muir's colleague, William E. Colby, enthusiastically echoed Muir's views on club outings. Colby's credentials as a conservationist were impeccable. He served as a club director for nearly half a century and became secretary of the club in 1900, a position he retained until 1946, except for two years when he was president.

Colby practiced law in San Francisco and also taught mining and water law at the University of California, Berkeley. When he was appointed the first chairman of the California State Park Commission in 1927, there were only five parks. When he left the commission in 1936, there were forty-nine state parks and eleven historical monuments.

After Muir died in 1914, Colby continued to preach the philosophy that "everyone should have an opportunity to see as much of the mountains as possible. Accordingly the Sierra Club proposed roads across Kearsarge and many other Sierra passes."[18]

Following World War II, this began to change with the next generation of leaders. In 1949, when David Brower, Richard Leonard, and several other directors "blocked a proposal for a road into the upper Kings River country in the Sierra," Colby resigned from the board in protest.[19]

The new leadership introduced the new philosophy of sharing "our" forestland. At "their suggestion the by-law instructing members to render the mountains 'accessible' was deleted."[20] The era of stewardship of our forestlands for a privileged few was launched.

The way the "privileged few" philosophy works is shown in the *New York Times* Buck Lake advertisement mentioned above, which was signed by the Sierra Club president. It opens with an attention-grabber in large print—"your forest"—followed by "your federal lands."[21]

This approach has become standard operating procedure for antiforestry advocates. They maintain that forestlands, both public and private, belong to all of us. But, having won over Congress and the courts, they have greatly expanded roadless designated reserves which are open to all wildlife but functionally closed to nearly all citizens.

Stewardship: Media Perceptions

The advertisements in the *New York Times* just reviewed are consistent with the paper's editorial and news coverage of the subject. It is difficult to understand what drives the editors of this newspaper to misconstrue the issues dealing with forestland conservation and management.

In the sixties, for example, readers were told, "they are cutting the last of the redwoods" and in the eighties, "they are cutting the last of the old growth."

Some knowledge of forestry is necessary to make such a statement as "they are cutting the last of the redwoods." The average person cannot visualize how many trees are in a forest stand. Let me try to explain. In the early days cruisers made their living by taking inventory on the stump; that is, tallying a sample of trees by species, volume, and grade on a preselected cruise line.

In 1953, I was one of three cruisers who carried out a 10 percent cruise on a 37,000 acre old-growth tract on Vancouver Island in British Columbia. Visualize square miles of forestland of 640 acres each which are further subdivided into 40s. A 10 percent inventory requires two cruise lines through a row of 40s across the square mile.

Now visualize tallying merchantable trees by size and grade on each cruise line. This was done by pacing five chains (330') in from the forest license boundary, running a one chain (66') wide cruise line, then off-setting 10 chains (660') and running another cruise line. A 40 (40 acres) is 1,320' wide. From daylight until near dark, seven days a week, three cruisers did this for ten weeks and we tallied the trees on only 10 percent of 37,000 acres.

Again, visualize 1.336 million acres of redwood forestland. You see the point I am making. To look at all of the redwood trees on both public and private land in California, it would take about three lifespans of a person "walking the woods," daylight to dark, every day of his or her life looking at redwood trees. That is, for a 100 percent cruise, it would take 208 years to see all the redwood trees.

Are we really "cutting the last of the redwoods"?

For several years, the *New York Times* has also been playing a leading role in confusing the old-growth issue in the Pacific Northwest. Nothing is more farfetched than asserting, "They are cutting the last of the old growth," when our designated reserves heavy to old growth would fill a land area equivalent to the states of New York and New Jersey combined.

Nonetheless, articles in the *NYT* continue to foment confrontation based on untruths. Evidently the editors are not interested in facts. For in another instance, they accept without question the notion that northern spotted owls require old growth to survive. This is a great hoax.

Do we know how many owls are in Pacific Northwest forests? Do we know how many there were ten years ago? One hundred years ago? Do we have accurate scientific data indicating that owl populations are declining? No to all these questions.

As seen earlier, the owl hoax was exposed by a representative of the Sierra Club Legal Defense Fund when he admitted in a public forum that the owl issue was fabricated to promote the real issue—to prevent harvesting on public forestland. Yet the *New York Times* continues to circulate the hoax.

On 26 December 1991, this newspaper printed an article under the headline, "Scientists Find No Evidence of Resurgence of Northern Spotted Owl." The article quoted a representative of the Fish and Wildlife Service as saying: "In terms of critical habitat, I've not seen new information that changes significantly our understanding of the biology of the Northern spotted owl." Therefore, government maps of critical habitat "will reflect the same scientific assumptions used since the bird was declared to be a threatened species in June 1990."

A few weeks later, on 9 January 1992, the *Times* stated, "Fewer than 500 of the birds are left in existence" and then assured that "no jobs would be lost." The very next day a New Jersey newspaper ran a more objective story informing readers that the U.S. Fish and Wildlife Service "predicted that overall efforts to save the owl will cost about 33,000 Northwest jobs."

When it comes to citing data related to the owl issue, government scientists and reporters are very selective, tending to favor data that are consistent with the preservationist ethic.

From the very beginning, confusion was inevitable because government research protocols for the owl census were designed to underestimate. A wildlife biologist in the private sector, Lowell Diller learned this in his first summer on the job in 1989. He determined that banding the owls was the only way to carry out a complete census and eliminate double counting. He and four members of his research team started banding in the summer of 1990, and through of 1993, over six hundred owls were banded.

On the basis of his work and that of other wildlife biologists in the private sector, credible scientific evidence points to a greater owl population in northern California than was thought to exist in all of California, Oregon, and Washington when the owl was listed as threatened.[22] Moreover, owls were commonly found in blends of second growth and clearcuts. They were not found in undisturbed old growth, so ardently defended by antiforestry advocates.

Following these findings, in October 1993 the California Forestry Association filed a petition with the U.S. Fish and Wildlife Service to delist the owl. Dr. Robert Taylor, the association's wildlife biologist, believes there is ample evidence to show that the owl should never have been listed as threatened in the first place.

The *New York Times* conservation policy poses a riddle. Why does a company that probably consumes more wood fiber per employee than any company in the world consistently denigrate professional forestry and the care of our renewable forestland ecosystems?

Stewardship: Economics and the American Dream

We have talked about wildfires and snags being killers. What about the killers of the American dream? At times, it seems, families who have attained the dream get more attention than those who have not. On 20 October 1991, a wildfire burned over two thousand acres, destroying more than 3,900 homes, townhouses, and apartment units in the Oakland Hills area, with a loss of twenty-six lives. Fire damages were estimated as over 5 billion dollars.

Naturally, this disaster attracted nationwide attention. But largely unnoticed, the American dream is also going up in smoke for millions who have never owned a home of their own—the dream is beyond their means, beyond their hopes.

These constitute the majority of American families. "The Census Bureau reports that 57% of American families can't afford a median-priced home near where they live."[23]

In 1991, a special Housing and Urban Development (HUD) commission comprised of builders, local government officials, and advocates for low-income housing addressed the question of afford-able housing. Their report cited the deleterious impact of the spot-ted owl and Steven's kangaroo rat. Their recommendations includ-ed "an overhaul of the Endangered Species Act."[24]

The report noted that "bureaucracy and red tape [have] added $15,000 to $30,000 to the cost of houses in many markets. . . .It should-n't surprise anyone then that nine out of 10 renters and three-quarters of Hispanic and black families are frozen out of the housing market."

The report then said that "the Steven's kangaroo rat recently became one of the largest 'landowners' in California when a 30-square mile stretch of land worth $100 million was declared off-limits to development in order to protect the rat." At this rate we are not far from a "Who's Who" of millionaire owls, woodpeckers, and rats.

"A Fish & Wildlife Service official," the report continued, "defended the move by saying that humans have reached the limit on how far they can intrude on the environment." The official's comments were accompanied by the familiar refrain, "I'm not required by law to analyze the housing-price aspect for the average Californian."

The American dream is tied closely to the Forest Products Industry. For fifty years, this industry has satisfied the ever-increas-ing demands of consumers. We have become the best housed nation in the world. But now, restrictive legislation is cutting away the resource base of the Forest Products Industry on public lands and raising havoc with raw material sources in the private sector as well. Homeownership statistics are telling us what to expect in the 1990s.

According to the Census Bureau, "Homeownership rates for people under age 35 dropped markedly in the past decade."[25] With few exceptions, young adults who wish to live in their home towns can no longer afford to do so. Statistics from 1980 through the sec-ond quarter of 1991 reveal just how far out of reach the American dream of homeownership has become for the younger genera-tion: Under age 25 it is only 15 percent, down from 21 percent; between ages 25 and 29, 34 percent, down from 43 percent, and from 30 to 34, 51 percent, down from 61 percent.

Stewardship: Economics and
the Inverted Pyramid

Fewer than one out of five gainfully employed members of our society is producing materials which enable us to live the good life. This person has to support not only himself or herself and family but also four others and their families: he or she is carrying the entire load of society at the apex of an upside-down pyramid. The other four out of five persons are working in government, secondary manufacturing, service jobs, or the fast-growing information sector.

By 1988, as an indicator of our society's shift away from manufacturing, government jobs surpassed manufacturing jobs. By 1990, there were a total of 21.1 million jobs in government compared to 19.7 million jobs in manufacturing, and the disparity keeps growing.[26]

Before life is crushed out of the remaining primary producers, the burden must be balanced and the pyramid put back on its base. Government and service jobs depend on primary manufacturing jobs. This is why the link between our renewable resources and primary manufacturing is so important.

The Forest Products Industry, a primary manufacturing industry, is as crucial as mining, steel, oil, gas, and other industries that are wholly dependent on our resources. Yet many stories give the misleading impression that primary manufacturing jobs can be readily replaced by opportunities in tourism and recreation.

But again, without primary manufacturing jobs, the service sector would soon collapse. Such a collapse would embrace wholesaling, retailing, health care, government employment, and all secondary manufacturing such as home furnishings, not to mention the burgeoning electronics industry.

Unless we can move in a positive, meaningful way toward solving the problems in regulatory gridlock, primary manufacturing will finally be crushed out of existence.

Stewardship: Economics and
the Forest Service Road System

A typical example of misinformation regarding the Forest Service road system appeared in a 9 December 1991 issue of *Forbes* titled, "Spare That Tree!" Five pages and six photos full of inaccuracies. The writer does not even bother to check the size of the system. Currently there are 369,000 miles of Forest Service roads, not the 342,000 reported in the article.

The writer subsequently indicates that the entire road system is made up of logging roads. In reality, logging activity comprises a small fraction of the road system traffic; recreation is the more common activity. Moreover, roads are essential for the administration and care of our Forest Service forestlands.

Nor is the road system invading Wilderness as the writer suggests. That, among other things, is against the law. Once again, as in many other articles, the Forest Service road system is compared to the Interstate Highway System; 342,000 miles compared to "some 50,000 miles." But it is yet another case of apples and oranges. In truth, not a single mile of Forest Service roads can be compared to a mile of the interstate system.

The Forest Service road system is part of the 3,921,000 mile road system in the United States. Here are some more statistics: In addition to the interstate system, which, incidentally, is 45,000 miles, there are 3,123,000 miles of rural roads and 754,000 miles of urban roads. Less than 15 percent of our nation's road system is unpaved, while over 98 percent of the Forest Service road system is unpaved.

Of the 369,000 mile Forest Service system, 75 percent or 276,000 miles are classified as "local" roads—single-lane, unpaved, with limited vehicular access. Twenty percent or 74,000 miles are considered "collector" roads—normally single-lane, gravel-surfaced, and all-weather roads. Five percent or 18,000 miles are "arterial" roads—usually two-lanes and about two-thirds paved.

Of the 369,000 miles, 66,000 are closed to the public; 210,000 are maintained for high-clearance vehicles, not recommended for passenger cars; and 93,000 are maintained for passenger cars.

Road systems are essential in order to enjoy the many recreational opportunities offered by the Forest Service System—to all of us. Furthermore, road access is essential so that government agencies can carry on their stewardship of public lands: recreation, wildfire detection and control, watershed protection, care of wildlife and plant habitat, access to raw material resources, and so on.

To say we should not have road access to forestlands, for whatever reason, is like preventing a driver from using roads to get to work, school, store, and church. With few exceptions, in our society the idea of not having road access is absurd.

Yet a steady stream of legislative initiatives are calling for more and more roadless areas. If appropriations over the past decade are any indication, members of Congress seem intent on putting the Forest Service out of the road-building business. In 1984, Forest Service road construction appropriations totalled $223 million. By FY 1993, these appropriations had dwindled to $141 million.

Stewardship: Economics and Below-cost Sales

Antiforestry advocates have devoted a good deal of time playing with a confusing array of cost figures which they maintain are not covered by Forest Service raw material sales. "Below-cost sales" is a contrived issue, nearly as confusing as the Forest Service road system, and the two are linked by antiforestry advocates.

The issue is—or should be—caring for our renewable resources. To suggest to the general public, as antiforestry advocates do, that a National Forest system sale must be profitable in the business sense is folly. Our public forestlands were never intended to be a profit-and-loss center any more than the United States Postal Service or Department of Defense.

Today we have the means to manage our renewable forestland ecosystems and contribute to their overall health. But this stewardship requires road access and forestland investment. Salvage sales should be routine even though they are "not profitable." Investment in our distressed forestland ecosystems should be a top priority for members of Congress; watershed protection, recreation facilities, and a healthy environment for our flora and fauna are not possible without investment.

Stewardship: Congress and Strategic Economics

During World War II, the Japanese considered our forestlands strategically important and tried hard to destroy them by fire. In 1942, incendiary bombs were dropped into southern Oregon forestlands, near Brookings and Cape Blanco, and in 1945, the Japanese launched thousands of large paper incendiary balloons into high air currents from their home islands, hoping they would drift across the Pacific and set our forests on fire.[27]

Some of these landed as far east as Michigan. One of the balloons caused the only wartime fatalities from enemy action in the continental United States when, in May 1945, six people were killed near Klamath Falls, Oregon. Much of this forestland is now known as "Owl Forestland," a part of our designated reserves. But is Congress aware of the strategic value of forestland stewardship? There is some evidence that citizens are finally becoming alarmed at the wrong turns Congress is taking. Primary producers are joining recreationists at the grass roots level to share their concern.

In March 1991, the Wilderness Impact Research Foundation held its fourth annual Wilderness Conference in Denver. Organizational cosponsors numbered over 230, representing some 25 million members. Several hundred attended the conference from twenty-eight states and five countries. The group seeks to convince the general public and members of Congress how short-sighted and counterproductive restrictive legislation has become.

One of the most telling presentations came from the late Rita Klimova, ambassador to the United States from the Czech and Slovak Federal Republic. She was astonished at how much control our federal government had amassed over the lives and properties of its citizens—especially shocking in view of the recent developments in Eastern Europe and the former Republics of the Soviet Union which are battling to erase central controls.

These people have experienced central controls and the stagnation it brings. And they have also experienced the immeasurable environmental costs to society brought about by central controls.

The implications of this global issue were clearly defined in June 1992 in Rio de Janeiro during the United Nations Conference on Environment and Development (UNCED). Here forty thousand environmentalists from all walks of life representing some 180 countries, along with eight thousand journalists and 121 heads of state, came together with the noble objective of saving the planet earth.

Two protocols dealing with global warming and biodiversity captured the attention of people around the world. President Bush received scathing criticism for signing only the one dealing with global warming. President Clinton signed the biodiversity document in December 1993.

The point here is that the general public seems unaware of the impact of these agreements on our ability to manage our renewable forestland ecosystems. When signed and ratified by the Senate, they carry the same obligation as a treaty and are enforceable by law. We are currently facing regulatory gridlock on a national scale. With these protocols, we face global bureaucratic mandates.

The remarkable advances in forestland conservation and management achieved over the past fifty years cannot survive central controls and global mandates either on a national or global scale.

Stewardship: Congress and Global Economics

So far, Congress has set aside some 95 million acres in our National Wilderness Preservation System. These Wilderness acres exceed the total land base of the home islands of Japan.

Federal set-asides are rich with resources—renewable forestland ecosystems, minerals, coal, oil, natural gas—all of which will remain in their natural state and, with few exceptions, inaccessible.

How does the prevailing concept of stewardship in Congress affect our ability to compete in global markets? And what about our trading partners on the other side of the world? Take, for example, the intensively managed forestlands in Sweden and Finland.

These countries are not only trading partners, but formidable competitors. Sweden's total land base is 101 million acres, 58 percent of which is forested. Finland's total land base is 83 million acres, 57 percent of which is forested. Our Wilderness Areas today exceed Finland's total area and will shortly exceed Sweden's.

It is unlikely that Sweden and Finland will set aside working commercial forestlands on a scale even remotely resembling that in the United States. It would be economic suicide. That being the case, our Forest Products Industry will be unable to compete globally.

If our essential daily needs do not come from our forestlands, where will they come from? Most children and a surprising number of adults view the local mall as the source. Aldo Leopold touched on this perception in A Sand County Almanac: "There are two spiritual dangers in not owning a farm. One is the danger of supposing that breakfast comes from the grocery, and the other that heat comes from the furnace."

And unlike our cavalier attitude to the source of our essential daily needs—our forestlands—our competitors from Sweden and Finland do not shun the responsibility of caring for their renewable resources.

Before restrictive legislation became the vogue, it was assumed by many forestland managers in the private sector, and even by some in the public, that potential and sustainable annual harvests from our national forests in the second half of the twentieth century would be in the 30 billion board feet (BBF) range.

But when we compare that expectation with the actual annual harvest from our National Forests—an average of some 12 BBF for the five years through FY 1990—it is easy to see that the charter of the Forest Service, "to furnish a continuous supply of timber for the use and necessities of citizens of the United States," has not been fulfilled.

If our public forestlands were being managed as well as those in Sweden and Finland, annual harvests would be in the 30 to 40 BBF range on a sustained-yield basis, an estimate some experts would term conservative. Instead, for fiscal year 1993, Forest Service sales were 4.5 BBF.

Consider this: When you divide 4.5 BBF by the 85.2 million acres of commercial forestland in our National Forest System, you get an average of 53 board feet (BF) per acre. This is less than ten pieces of 6 foot pine shelving from your local lumberyard. The commercial value of raw material from our National Forests is close to zero!

What is the *potential?* No one really knows. (*Potential* should not be confused with yield per acre or actual harvest per acre.) But let's apply a modest volume per acre, say 7,500 BF, which would give a *potential* of 639 BBF. Sensible stewardship in the private sector results in five to ten times this volume per acre. You see the point. The specter of overcutting is nothing more than a myth. The recent annual National Forest harvests of 12 BBF are less than 2 percent of a conservative potential and a 30 BBF annual sustainable harvest would be less that 5 percent!

Our forestland blessings are so striking and the potential so vast that the results of years of tinkering by antiforestry advocates have only recently translated into hardships for society at large. But finally the dereliction of our responsibilities is taking its toll.

Stewardship: Congress and Domestic Economics

Let us look at what regulatory gridlock does for domestic economics. We will use floor joist material as an indicator (i.e., kiln-dried Douglas-fir, #2 & Better, 2X10, random lengths, in carlots, FOB mill). Almost every single-family residence with suspended wood floors built since World War II utilizes a quantity of this 2X10 lumber.

In 1991, according to *Random Lengths*, a weekly market newsletter published in Eugene, Oregon, raw-material shortages in the Far West pushed the carlot price of this item from a low of $255 per thousand board feet (MBF) in January to a high of $465/MBF in June.[28] The cost of replacing inventories of joist material in many retail lumberyards was doubled.

This hike in the price of framing lumber occurred when the home building industry was in a depression, with annual housing starts barely 1 million units, the lowest in forty-six years. Worse comes when we move into a positive business cycle. In March 1992 and again in March 1993, *Random Lengths* noted that commodity lumber prices were at all-time highs. Floor-joist lumber topped out at $635/MBF.

The last boom years in housing construction were in the 1970s, when starts exceeded 2 million units per year in 1971, 1972, and 1973, and again in 1977 and 1978. Congressional hearings were held during both of these boom periods to investigate the high prices.

Forest Products Industry representatives were called to appear and they explained the basic law of supply and demand. Yet simultaneously, more restrictive legislation, initiated by some of these same members of Congress, was already being proposed, further limiting the ability of the industry to come up with an adequate supply of framing lumber, as well as other essential needs of the nation.

The mill price of floor joist material continues to rise of course (two-and-a-half times in twenty-four months); the next shock will be what is called "stock-outs"—no framing lumber at any price. Media people like the phrase "long lines and empty shelves" to describe this misfortune. At this rate, the American dream of home-ownership will be completely out of reach.

STEWARDSHIP: SHARING CONCERN IN A POSITIVE WAY

All this is not to say that we do not need environmental measures. We do, but they must be forthright and effective. Indeed, we would be in more trouble than we are without them.

We also need research in both public and private forestlands. For example, the Olympic Natural Resources Center, soon to open in Forks, Washington, is a giant step in carrying out forestland research on a realistic scale.

Our National Park System is also necessary. But we should not turn vast tracts of our National Forest System into designated reserves, which is tantamount to turning them over to the National Park System. These two systems were chartered to serve two different purposes, a distinction largely overlooked by Congress.

A close watch should also be kept on the Forest Products Industry to ensure proper forestland stewardship. Remember, there are only a few hundred major forestland owners in the industrial sector and some 6 million owners in the nonindustrial sector. We should not confuse them.

Other worthwhile programs include the National Wilderness Preservation System, the Wild and Scenic Rivers System, the National Wildlife Refuge System, and the National Seashores and Lakeshores System. All these systems set aside land and water for our citizens to enjoy.

What is harmful is the setting aside of tens of millions of acres of fully stocked forestlands in roadless designated reserves which exclude ecosystem management.

Assuring positive initiatives for stewardship will be an uphill battle. This was evident in a January 1992 workshop in Reston, Virginia, which sought to define sustainable forestry. Some 150 people were invited.

The background statement read in part: "The emphasis on timber yields is derived from a 19th century paradigm that holds all other forest values as compatible with timber production. After 100 years, however, it is clear that we are no closer to sustainable forest ecosystems today than we were when the Organic Act was passed in 1897." That, to put it kindly, is misleading.

Sustainable forestry is what professional forestry is all about. Industrial and nonindustrial forestry are compatible with watershed protection, recreation, and proper habitat for wildlife and plants. Professional forestry does not drive any species to extinction. On the contrary, it helps them thrive. Moreover, professional forestry and biodiversity go hand in hand. In reality there is common ground.

But during the working group discussions, in answer to a leading question, "What do we need more of and what do we need less of for sustainable forestry?" a large number of participants seemed to agree that we needed fewer commodities. This is like saying we need less air to breathe, less water to drink, and less food to eat.

The passion and care exhibited by today's professional foresters for our renewable resources belies the mindset that laid waste the fabled cedars of Lebanon and turned the Middle East and much of North Africa into a desert. Professional foresters renew and manage our forestland ecosystems on a scale never before seen in history.

But there is hope. In November 1990, some thirty people gathered at Grey Towers, the family home of Gifford Pinchot in Milford,

Pennsylvania, for a two-day conference on forestland stewardship. From this session came "The Grey Towers Protocol," which is summarized in the booklet, *Land Stewardship in the Next Era of Conservation*, by V. Alaric Sample.

This is part of the Breaking New Ground series of the Pinchot Institute for Conservation, located at the Grey Towers National Historic Landmark. The Grey Towers Protocol ought to be read by all forestland practitioners, both industrial and nonindustrial. The four principles are:

1. Management activities must be within the physical and biological capabilities of the land, based upon comprehensive, up-to-date resource information and a thorough scientific understanding of the ecosystem's functioning and response.
2. The intent of management, as well as monitoring and reporting, should be making progress toward desired future resource conditions, not on achieving specific near-term resource output targets.
3. Stewardship means passing the land and resources—including intact, functioning forest ecosystems—to the next generation in better condition than they were found.
4. Land stewardship must be more than good "scientific management"; it must be a moral imperative.[29]

FORESTLAND STEWARDSHIP: THE PRIVATE SECTOR

Much has been accomplished to put the principles outlined in the Grey Towers Protocol into effect by industrial and nonindustrial forestry programs over the past fifty years. Of course, much work remains. But we are moving in the right direction.

The photos in this section show views of forestlands, primarily in the Far West, where I worked in my younger days. Most of the photos were taken in July 1991 and cover an area from the northern end of Vancouver Island to the Calaveras Big Trees State Park in the Sierra, between Lake Tahoe and Yosemite.

MacMillan Bloedel and Canadian Forest Products, Canada

The ten-week cruise mentioned earlier was in a remote area near Port McNeill on British Columbia's Vancouver Island. In 1953, after a commercial flight from Vancouver to Alert Bay, a bush pilot dropped us onto Maynard Lake (19 miles inland from Port McNeill). We were miles from the nearest settlement and, until the bush plane picked us up, we never saw anyone except members of our eight-man work party.

This was indeed old-growth forestland, overmature since the turn of the century and beyond its prime for forest products. Thirty-six years passed before I was able to revisit these forestlands, now managed by MacMillan Bloedel and Canadian Forest Products.

Today the entire area has access roads which permit a great variety of recreational pursuits. Well over half of the tract has been logged, which started shortly after our cruise. The vigorous, healthy second growth which is replacing the old growth is as beautiful as anything you can imagine. The makeup of the new forests in terms of species mix and density is similar to the original forests.

If this renewal had followed a catastrophic wildfire or volcanic eruption, delighted scientists would be rushing to the scene, rejoicing at the beauty of nature and the forestland's ability to renew itself. But no plaudits go to managed renewal that achieves the same desirable ends—only criticism.

Habitat diversity for both wildlife and plant life is unmatched. Water quality has never been better. Finally, and perhaps most significantly, recreation within these managed forestland ecosystems is flourishing. Unlike 1953, when only a select few could gain access, today thousands of Canadian and United States visitors are able to take advantage of recreational opportunities.

The road systems are excellent and MacMillan Bloedel provides detailed maps of the region. There is much similarity between these Canadian managed forestlands and Bavarian forestlands. For information beyond the photo essay, I recommend a fourteen-page brochure, "Future Forests," published in November 1990 by MacMillan Bloedel (Box 6300, Vancouver, B.C., Canada, V6B 4B5).

1

2

3

4

5

6

7

8

9

10

11

12

13

14

15

16

17

18

19

20

21

22

23

24

25

26

27

28

29

30

31

32

33

34

35

36

37

38

39

40

41

42

43

44

45

Photo Essay: The forestlands we cruised stretch from Maynard Lake (our first base camp in 1953) south and east across the northern tip of Vancouver Island to Atluck Lake. These photos give an on-the-ground look at how renewable resources are nurtured in the real world:

- (Photo 13): We are well above Maynard Lake, which is off the picture to the right, and looking west past Benson Lake, in the middle of the photo, and Kathleen Lake beyond. Port Alice can be found through the opening in the high country in the background.

    A good deal of old growth has been removed, with the most recent harvesting clearly visible in the foreground. Old growth is scattered throughout, however, from the patch in the lower left corner to the areas around the lakes and the very large tracts in the background. This area has not been deforested and young trees are conspicuous in the background and in the left foreground.

- (Photo 14): We are looking south from high up in the Maynard Lake block. In the foreground is Iron Lake. Beyond and off to the left is Lac Truite, our second base camp in 1953. The snow-covered high country on the skyline is Snow Saddle Mountain. Merry Widow Mountain is off the photo to the right, but mining activity at its base is evident in the distance. While the main road is just visible on the near side of Iron Lake, as is a secondary haul road on the far side, most of the roads in this photo are obscured by a healthy stand of second growth, most conspicuous in the foreground.

    Harvesting plans call for leaving areas of old growth around the lakes and in scattered areas along the drainages. Old growth is abundant in the higher elevations. Just beyond Lac Truite on the edge of the photo is the preferred home of the resident elk herd. They bed down in this area of old growth and then go out into the harvested areas for their "groceries." Three of the elk are monitored by MacMillan Bloedel biologists. The history of this herd is being studied closely by these scientists.

There were no elk here in 1953. They have been attracted by browse which is a result of harvesting. If old growth were essential habitat for elk, our eight-man crew would have encountered them on a regular basis. This was not the case. The animals and their sign were not to be seen. Remember, we were covering a lot of territory, across drainages, around lakes, with some cruise lines to the ridge tops and we were moving quietly.

One of the more prevalent and misleading notions is that wildlife and plants are more abundant in old growth than in managed forestland ecosystems. There is no reliable data to support this contention. In fact, where serious research has been undertaken, the reverse seems to be the case. This is no recent finding.

George Perkins Marsh, scholar, statesman, diplomat, and author, had this to say in the nineteenth century:

In a region absolutely covered with trees, human life could not long be sustained, for want of animal and vegetable food. The depths of the forest seldom furnish either bulb or fruit suited to the nourishment of man; and the fowls and beasts on which he feeds are scarcely seen except upon the margin of the wood, for here only grow the shrubs and grasses, and here only are found the seeds and insects, which form the sustenance of the noncarnivorous birds and quadrupeds.[30]

Marsh also cited Dr. Newberry's 1857 Report on Botany:

Dr. Newberry, describing the vast forests of the yellow pine of the West, Pinus ponderosa, remarks: "In the arid and desert regions of the interior basin, we made whole days' marches in forests of yellow pine, of which neither the monotony was broken by other forms of vegetation, nor its stillness by the flutter of a bird or the hum of an insect."[31]

- (Photo 15): We are looking southwest toward Merry Widow Mountain (snow covered on the left skyline), with Iron Lake running north and south in the middle. We are still in the Maynard Lake block viewing a large area of forestland which we cruised in 1953. Beauty is, indeed, in the eye of the beholder.

  This area is blessed with excellent soils, climate, and species for managing renewable forestland ecosystems. Note the vigorous new forest in the foreground, old growth around the shore of Iron Lake and beyond to the left, and considerable old growth in the high country across the skyline. Recent harvesting is conspicuous; but just as conspicuous is the vigorous second growth between Iron Lake and the old growth on Merry Widow Mountain.

- (Photo 16): This is a close-up of the drainage between Maynard Lake (out of the photo on the right) and Iron Lake (off to the left). Here you can plainly see the contrast between old growth in the background and left foreground and recent harvests. Running through the recently harvested areas is a broad band of vigorous second growth where old growth was harvested some twenty years ago.

- (Photo 17): This close-up view shows the contrast between old growth and second growth. Barely visible in the upper right is a portion of Iron Lake. The old-growth forest in this Maynard Lake block is about 50 percent hemlock and 25 percent balsam fir, with the balance mostly red cedar intermixed with spruce, yellow cedar, and Douglas-fir. The new forest is returning in much the same species mix and density as the old growth which it replaced.

- (Photo 18): This close-up of second growth along the main road near Port McNeill was harvested before our cruise in 1953. It is all natural regeneration. Here again the new forest is taking on the characteristics of the earlier old growth, although some scientists would say this is not a forest.

- (Photo 19): And here is a close-up within the second-growth forest of the previous photo. Happily, thanks to the intensive research conducted within the Mount St. Helens National Volcanic Monument, we know considerably more now about what makes up critical habitat and how difficult it is to destroy.

  Scientists found over one thousand species in a hostile environment within the blast zone of Mount St. Helens. When looking at this photo, one can safely assume that the same intensive research would reveal even more species.

- (Photo 20): We are looking south from the end of the MacMillan Bloedel road system. The drainage in the foreground flows around our viewpoint and north to Charlotte Strait. The drainage system off Mount Renwick, the snow-covered peak on the skyline, flows south toward Tahish Inlet and the Pacific Ocean.

  The high forested ridge in the left background is old growth. It is difficult to see the dead, dying, and down trees in the photo, though an indicator is visible in the remnant of old growth on the skyline at the right. The dead outnumber the living in this remnant. In the foreground is a healthy new forest, mostly hemlock, and in the background a more recently harvested section, which will look like the foreground in a few years.

- (Photo 21): This forestland is managed by Canadian Forest Products. We are above Hustan Lake (off the photo to the right) and looking due south at Pinder Peak (elev. 5,059'). Atluck Lake (the site of our final base camp) is beyond the old-growth forest in the middle of the photo at the foot of Pinder Peak.

Hustan Lake is midway between Atluck and Nimpkish Lake. The relatively new main North Island Highway 19 runs north and south along Nimpkish Lake for 15 miles. (This area was roadless in 1953). The only harvested area visible in this photo is the foreground. The rest of the view is as it was in 1953. We have more Douglas-fir in the new forest; note the healthy growth on the left, about a foot and a half each year.

- (Photo 22): Here is a close-up of Atluck Lake and Pinder Peak. No harvest; all old growth. The large alluvial fan in the center is an open meadow formed by runoff from Pinder Peak.

- (Photo 23): Another close-up of Atluck Lake, with Rita Lowell and our granddaughter, Sarah, important members of our photography crew. You are looking at the beauty of road systems for family recreation.

- (Photo 24): This is the northeast end of Atluck Lake—the opposite end—five miles from our final base camp in 1953. It is midweek and there is little activity. But on weekends and holidays, recreationists of every stripe converge here to enjoy this pristine location maintained by Canadian Forest Products.

  In 1953, it was functionally closed to citizens as are our Wilderness Areas today. Nearly all of the forestland in this photo is old growth except for the recently harvested open area, just greening up with young trees. Note the healthy new forest in front of the open area on the right, replacing a small tract of old growth harvested shortly after our cruise.

- (Photo 25): This view is a short distance around the lake from the previous view, with part of the recreational area on the left shoreline. Again, one can observe the contrast between the healthy new forest in the middle and the surrounding old growth.

These photos show the sharp contrast between ecosystem management and leaving these forestlands to "nature." For professional foresters who have dedicated their lives to nurturing forestlands, nothing is more depressing than close-up views of dying, dead, and down trees. Forestland health and wildlife habitat in these photos are light years better than in 1953. Now, surrounded by a vibrant, vital ecosystem, the sounds of life, animals, and birds, are a welcome addition to the sounds of the wind and the rain and the rush of streams.

Weyerhaeuser Company: Biodiversity

Are complex ecosystems and biodiversity compatible with multiple-use? Antiforestry advocates say not. As for distinguished scientists and educators, one educator will extol the virtues of "new forestry" vs. "cornfield forestry," while another will advocate "multiresource forest management" vs. "plantation forestry." Is it possible to obtain a clear answer? The eruption of Mount St. Helens in 1980 sheds light on these issues in several interesting ways. Today, on Weyerhaeuser's property along the Green River within the blast zone, there stands a lush new forest with trees higher than two-story buildings. (This is an area where I worked while a student in the late 1940s and early 1950s.)

The biodiversity in Weyerhaeuser's St. Helens Tree Farm is closely monitored by an environmental science research team which includes aquatic and wildlife biologists, hydrologists, silviculture and soil scientists, as well as professional foresters. So far, within this renewed forestland ecosystem, this research team has identified some 125 species of plant life and nearly 150 species of wildlife.[32] Here we have ecosystem management at its best. And yet this showcase is maligned as being nothing more than a "monoculture" or "cornfield forestry."

There are other positive features worth noting on Weyerhaeuser's tree farm within the blast zone. The entire area is well roaded for access by employees and visitors. And there are no signs of erosion from either the massive salvage logging operations after the blast or from construction of the extensive road system.

The quality of water found here enhances fisheries and dispels the notion that harvesting destroys watersheds and fisheries. Actually, thanks to exemplary wildlife management, hunting has never been better and fish counts are higher now than before the eruption.

For those interested in visiting this area, stop by the Weyerhaeuser office in Longview for copies of the St. Helens West (1990-91) Hunting Map published by the Washington Forest Protection Association, courtesy of Weyerhaeuser Company, Longview Fibre Company, Department of Resources and Department of Game. These maps not only orient the visitor, they also indicate numerous areas within the tree farm where deer and elk have overpopulated their own habitat. Obviously these animals prefer Weyerhaeuser's managed forestland to designated reserves where there is little forage.

Not many readers will have the opportunity to see the St. Helens Tree Farm firsthand, but Weyerhaeuser has made available a fifteen-minute video titled ". . . the trees go on forever" (May 1991), which anyone can request by writing to Communications Director, Weyerhaeuser Company, Tacoma, WA 98477.

I would like to share with you here a few photos of Weyerhaeuser's professional forestry prowess, information about the American Tree Farm System, and an historical note about Dr. Carl Alwin Schenck, the founder of the first school of forestry in America.

- (Photo 26): We are in southwest Washington and I have just taken this picture of our photographer, Mike Lowell. We are on a Weyerhaeuser haul road in the middle of the St. Helens Tree Farm about midway between Hemlock Pass to the south and the South Fork of the Toutle River to the north. Here is still another example of ecosystem management. This new forest is teeming with biodiversity. Actually, there is no such thing as a "monoculture" in our forestlands, whether they are natural or managed. There are hundreds of species in this photo. Plantlife is thick and heavy; it is difficult to walk through this thicket.

  For those interested in seeing these renewed forestlands in their early stages, pick up a copy of the Hunting Map mentioned earlier and take Interstate 5 northbound from Longview and Kelso. Go 6 miles, turn right, and head east for about 6 miles on Headquarters Road. At Headquarters, take the 1600 line road east and slightly south for another 6 miles to Hemlock Pass.

From here you can get a good panoramic view of ecosystem management. Backtrack a few hundred yards to the 4700 line and head north to the South Fork of the Toutle River, about 7 miles. Follow the South Fork west to the old Spirit Lake Highway (state route 504) which will return you to Interstate 5 at Castle Rock, some 10 miles north of Longview and Kelso.

A note of caution to readers who visit this country. Be sure to receive permission from Weyerhaeuser to use their private road system, and be alert at all times; these are working commercial forestlands.

• (Photo 27): This huge stump tells a story from the nineteenth century. In the photo we are looking north on the way to Hemlock Pass from Headquarters, and Mike Lowell is reading the story from springboard notches. In the foreground, you can see that a falling crew had left this area hours before we arrived. A rigging crew, loaders, and drivers will follow shortly to load and transport the logs to the mill.

About a hundred years ago, two fallers, each using a pair of springboards, notched their way up from the ground. (The third notch is by Mike's hand). When they got high enough to see where they wanted to fell the tree, they proceeded with the undercut using double-bitted axes. They then brought the tree down with a long two-man falling saw.

In order to cut through a Douglas-fir of this size, they used their springboards some 12 feet off the ground to walk around the trunk with a series of notches. Fascinating to follow their expertise in those days. Even more fascinating to realize that a whole new forest has again been harvested since those two lumberjacks left the scene.

Yet the perception persists that some hundred years ago the old-growth forest was "destroyed forever." This was about the time the Weyerhaeuser Company invested in the Far West. The reason is best answered by the oft-quoted statement, attributed to the company's founder, Frederick Weyerhaeuser: "This is not for us, nor for our children. It is for our grandchildren."[33]

The work and vision of this farsighted man was carried on by his grandson, Phil Weyerhaeuser, who made the momentous business decisions to end the cut-out-and-get-out policy and instead invest huge sums of money in harvesting and growing. That is the reality you can see in this photo.

After the original old-growth forest was harvested, a new crop of Douglas-fir sprang up from natural regeneration. Before the 1991 harvest, this new forest was taking on the composition, structure, and function of the original old growth. Note the old-growth bark characteristics of the stump in the foreground.

The 1991 harvest is the first step in the regeneration of the next forest. It, too, will resemble the nineteenth-century old-growth forest, but it will be accelerated—replanted within a year with genetically improved, hand-planted seedlings, and helped along by scientific and technological advances of recent years. And by 1995, all of the stumps in this photo will be obscured by a vigorous new forest.

American Tree Farm System

Weyerhaeuser's managed forestlands, while extensive, are only a small part of the American Tree Farm System, whose fiftieth anniversary was celebrated in 1991. Weyerhaeuser dedicated the first official Tree Farm on 12 June 1941, a 120,000 acre tract of logged-off, burned-over land in Grays Harbor County, Washington.

Tree farms, one of the most exciting concepts in ecosystem management, are little known by the general public. For an excellent write-up, the reader can request from Weyerhaeuser a copy of its six-teen-page brochure, A *Celebration for Generations to Come, Fifty Years, Clemons Tree Farm, 1941-1991* (August 1990).

Governor Arthur B. Langlie, in his dedication speech, speculated: "The Clemons Tree Farm may form the experimental basis that may mean a great deal to the entire state. It may set the pace for millions of acres of such lands throughout the state."[34] As it turned out, there was more vision than speculation in the governor's words.

The Clemons Tree Farm now comprises 400,000 acres. In addition, there are more than nine hundred dedicated tree farms in Washington, covering some 5.6 million acres of both industrial and nonindustrial forestlands. Moreover, tree farms have sprung up in every state, more than seventy thousand in all, with some 93 million acres under management.

Of course, there is considerable variance in the management of the individual tree farms due to climate, species, and soils which vary widely from one region to another. Also, as expected, large industrial tree farms have management goals which differ from small nonindustrial tree farms.

But various patterns of management—both commercial and noncommercial—provide benefits to society at large. While many tree farmers have little or no interest in turning a profit, the majority look to sales of logs, Christmas trees, floral greens, and fuel wood for income. Whether management goals are commercial or not, a portion of the income is nearly always reinvested in the tree farm. Income is often critical for retirement or for the education of children and grandchildren.

When historians start to research "ecosystem management," they will be enlightened by visiting tree farms. For tree farmers have long been concerned with healthy watersheds, wildlife sanctuaries, recreational opportunities, and converting dying, dead, and down forests that give off carbon dioxide into vibrant, healthy forestland ecosystems that give off oxygen.

Researchers will also find tree farmers who are only interested in providing a human sanctuary from the hubbub of daily life, whether in a remote area or within a few miles of urban centers.

Researchers will also learn that spiritual values and commercial values in our forestland ecosystems are not mutually exclusive. Common ground is there to be embraced.

Weyerhaeuser: Oregon Millicoma Tree Farm

- (Photo 28): The Millicoma Tree Farm is near Coos Bay on the southern Oregon coast, an area where I worked first as a cruiser and then with the Weyerhaeuser Company. This photo, courtesy of the Weyerhaeuser Company archives, shows the 7 July 1951 dedication of the 20,000 acre Carl Alwin Schenck Working Circle located between the Millicoma and South Coos Rivers. This picture shows one of the most extraordinary men in the annals of professional forestry in the United States. The view is looking east and a little north from McKeever Lookout. The lookout was dedicated that day to Don McKeever, a Weyerhaeuser research director, who was killed in a snag-felling accident in 1949. His widow, Mary, and their two boys, Dave and Don, are on the platform beside Dr. Schenck.

  Just beyond the platform is a portion of the Schenck Working Circle, primarily young growth. This forest replaced one harvested just prior to World War I. Note the snags from the original forest standing well above the new forest like lightning rods. Removal of these started shortly after the dedication.

  The Weyerhaeuser tribute to Dr. Schenck was part of a grand tour he made in 1951 visiting all the forested regions in the United States and Canada. Graduates of the Biltmore Forest School (described in Chapter 7) gathered around him at every stop with an outpouring of affection and appreciation.

- (Photo 29): This is the 1951 dedication scene as it appeared in July 1991. Only a few scraps of lumber remain from the platform. McKeever Lookout has long since been abandoned and the site is now known as McKeever Point. The 1951 view is obscured by a new forest; Jim Clarke, Weyerhaeuser forestry manager, standing at the right, provides perspective for the size of the trees. The haul road along the top of the ridge, which ran by McKeever Lookout, has been closed until the next harvest.

- (Photo 30): This view was taken within the second growth across the roadway from McKeever Point. Is this biodiversity?

Longview Fibre Company:
Exemplary Forestland Stewardship

This large Forest Products Industry company headquartered in Longview, Washington, provides another excellent example of renewable forestland ecosystem management in the private sector. Its spring 1992 *Longfibre News* gives six case histories of decisions that have greatly benefited citizens who live near their forestland holdings in Oregon and Washington.

These wide-ranging decisions highlight cooperative programs with federal, state, and local agencies, along with various tree farm neighbors, including recreational and environmental groups. Involved were a number of value-for-value forestland exchanges to enhance fisheries and wildlife habitat as well as recreational sites. Also involved were cooperative projects on Longview Fibre forestlands.

One of the most compelling studies is case study four, which describes how prime spawning grounds were enhanced for salmon and steelhead on the West Fork of Oregon's Hood River, which flows for several miles through the heart of a 20,000 acre portion of Longview Fibre's Mid-Columbia Tree Farm.

The virtual blockage of fish attempting to migrate upstream in this drainage had not been caused by human activity, but by the natural erosion of waterfalls which grew to 14 feet, nearly insurmountable for migrating fish (see photo 31).

The Bonneville Power Administration, working closely with the Oregon Department of Fish and Wildlife, retained the Portland geotechnical engineering firm of RZA AGRA, Inc., to develop a solution—the "moving-falls fishway." This unique design featured a series of concrete weirs extending 120 feet along the streambed and spanning the entire river channel, nearly 100 feet.

The new fishway is much larger than average fish ladders, making it possible for returning salmon and steelhead to migrate easily upstream beyond the high waterfall. Photos 32 and 33 show the waterflow over the concrete weirs before and after.

Willamette Industries:
Multiple-Use and Watershed Protection

One of the best multiple-use case histories in the private sector appeared in the 1988 Annual Report of Willamette Industries. This forestland conservation and management story was so well received it was reprinted the next year as a six-page fold-out.[35]

Regarding watershed protection, in the mid-1960s McMinnville, Oregon (population 16,000), developed plans for a municipal water supply reservoir and found that Willamette's forestlands offered the best and most economical site. The McMinnville experience clearly indicates that watersheds and industrial forestry are compatible.

Resident forester Dan Upton was honored by the Oregon State Department of Forestry with a Merit Award for exemplary management of the Trask Mountain Tree Farm, which includes 97 percent of the shoreline of McMinneville's reservoir. The Oregon State Department of Forestry also gave its Western Oregon Operator of the Year award to Willamette's major logging contractors. This award was for sensitive harvesting operations along Teal Creek in Willamette's Black Rock Tree Farm just above the intake pipe for the Falls City municipal water supply.

Those who wish greater detail may request a copy of the folder *Resource Stewardship at Willamette Industries, Inc.* from the manager of Corporate Communications in Portland, Oregon. This folder describes forestland stewardship at its best, now commonly referred to as ecosystem management.

Willamette Industries is also a major forestland owner in the southern pine region, where its wildlife management efforts have been outstanding. In Louisiana, they manage 563,000 acres of pine and hardwood forestland.

A significant portion of the Jackson-Bienville Game Management Unit of the Louisiana Department of Wildlife and Fisheries falls within the boundaries of Willamette Industries' forestlands. Over the past thirty years, multiple-use has gone hand-in-hand with the enhancement of wildlife and fisheries habitat to make this one of the most popular recreational areas in northern Louisiana, living up to the state's license plate motto, "Sportsman's Paradise."

Despite this outstanding record of stewardship, Willamette has been forced to abandon a number of its operations in Oregon because expanding designated reserves have eliminated its source of raw material.

Pacific Lumber:
A Positive Case History Cloaked in Darkness

In the previous chapter, the *National Geographic* alluded to a "cut-and-get" policy by Pacific Lumber. But quite the contrary, Pacific Lumber, in the redwood business since 1882, was one of the first operators to adopt a policy of sustained-yield—the forerunner of ecosystem management. Much inflammatory propaganda notwithstanding, this policy is still in force. Nor has this policy prevented Pacific from working closely with the Save-the-Redwoods League.

From 1931 through 1986, Pacific Lumber has given or sold to the state of California some 20,000 acres of redwood forestlands for parks and highways. These parcels vary from a few acres to 9,400 acres in size and include the well-known Avenue of the Giants in Humboldt Redwoods State Park.

Throughout, and particularly over the past twenty years, Pacific has continued to invest in redwood forestland and mills. Major changes include expanding from two sawmills in Scotia with seven hundred employees in 1970 to four sawmils with 1,300 employees, having added one in Fortuna (1972) and the other in Carlotta (1986). During this same period, forestland holdings increased from

165,000 to 194,000 acres.

Moreover, the annual harvest doubled from around 175 million board feet in 1985 to some 350 million in 1990. Antiforestry advocates, supported by numerous media reports, contend that the harvest of old-growth redwood has also doubled. But Mill B, Pacific Lumber's large-log redwood mill in Scotia, has held to a one-shift work schedule and an unchanging production output. And Mill A, the small-log mill in Scotia, and the mills in Fortuna and Carlotta have increased their output by utilizing small logs, not old growth.

The facts: Since 1986, the yearly harvest of forestland has increased about 15 percent from some 5,500 acres/year to 6,300. This is in line with the 18 percent increase in total forestland acreage under management since 1970. But the yearly harvest of old-growth redwood has actually decreased from some nine hundred to eight hundred acres.

The reason Pacific Lumber has been able to double the annual harvest with only a modest increase in acres harvested is because of the high yield/acre in both young growth and residuals. The latter are stands which have been commercially thinned under Pacific's sustained-yield management.

Inventory figures provided to management by an outside forestland consulting firm in 1956 and again in 1986 indicated that volumes, almost entirely in second growth, were 30 percent higher than was being carried on the books. Thus, Pacific Lumber can look for further increases in production, but again, without any increase in old-growth harvests.

Pacific Lumber: Shedding a Little Light

Returning to the photo essay: After leaving Weyerhaeuser's Millicoma Tree Farm and the coastal Douglas-fir forestlands of Coos County in southwestern Oregon, we continued into the redwood/Douglas-fir forestlands of Pacific Lumber in northern California.

Because of the controversy surrounding Pacific Lumber, I wanted a first-hand look. This region was familiar as I had worked nearby on a cruising assignment in the summer of 1952 as a compassman and mapper.

The photos included are near Shively Road, some five miles

south of Scotia along the Eel River. Scotia, the headquarters of Pacific Lumber, is 22 miles south of Eureka on US 101. The photos show a residual stand of second-growth redwood that was commercially thinned in 1965 after a clearcut of a century ago.

Not only Pacific Lumber, but all Forest Products Industry operators in northern California are fighting a losing battle against various antiforestry groups, which again are supported by a large proportion of the media and politicians.

Nearly a quarter of a million acres of redwood forestlands or about 18 percent of the total (1.336 million) have already been set aside as designated reserves in state and national parks and other public ownerships. These set-asides include some 76,000 acres of old-growth redwood.

Yet it has become next to impossible for operators in the private sector to carry out sensible stewardship of the remaining redwood forestlands. State politicians have done an excellent job of positioning themselves between redwood forestlands and licensed professional foresters.

- (Photo 34): This view is just off Shively Road a few hundred feet in a Pacific Lumber redwood forest. Note the high stump in the background from the earlier clearcut and the healthy stand of residuals all around. I am counting the rings on a recently felled tree and have determined its age to be somewhere between ninety and one hundred years.

  This redwood tree was half again to twice as large as a two-hundred-year-old tree in the Muir Woods National Monument 17 miles north of San Francisco. The marker on the Muir Woods tree reads something like, "This redwood tree was growing when the Declaration of Independence was signed."

  But in a century, on working commercial forestlands, we can create redwood trees as large as these two-hundred-year- old retired redwoods in the parks. (Are we in awe of them because of their age or their mere existence?)

- (Photo 35): The cameraman has pivoted 180 degrees and

is now looking north. This photo features a high stump from the earlier clearcut in the center background, a fair-sized stump sprout on the left (about three and a half feet in diameter), and a young redwood seedling in the foreground.

One can also get a pretty fair idea of how Pacific Lumber's sustained-yield management program looks on the ground, where only a small number of mature trees are harvested each year, leaving a great number of residuals for future harvests.

• (Photo 36): We are now looking west, into the canopy over Shively Road. Are we viewing a renewable resource?

Boise Cascade Corporation: Forestland Health

Nature is not kind to trees and forestlands. Nature is not kind to our environment. Long before civilization, our planet endured cycles of every imaginable trauma. During these upheavals, both animal life and plant life were often completely eliminated. To this day, we witness wildfires, volcanic eruptions, hurricanes, ice storms, insects, disease, and drought. By and large, the stress which nature places on forestland ecosystems makes any human stress look puny by comparison. And humans have done and are continuing to do something about it.

In the fall of 1992, Donna and I traveled through western Montana, Idaho, and the Blue Mountains of eastern Oregon to inspect the "forest health emergency" in our national forests. Just north of Boise, Idaho, the mess we observed was in stark contrast to a beautiful sight right alongside. The Boise National Forest, a sick ecosystem defaulted to nature, is a vivid testimonial to regulatory gridlock. Beside it is the Boise Cascade Corporation, a healthy, thriving managed ecosystem, a reassuring testimonial to ecosystem management, one of the most inspiring sights found anywhere.

Forestland management prescriptions in the dry Intermountain region are as different as night and day from those in the temperate

marine climate along the Pacific Ocean. But professional management can have the same results. No clearcutting here. No soil erosion. Excellent watershed protection. Beautiful, healthy, vigorous
stands of mixed species and different age classes.

The risk of wildfire is there, but there are no fuel loads such as are
found in adjacent unmanaged forestlands in the public sector. Plant and
wildlife habitat is top quality, and a wide range of recreational pursuits
are encouraged. Naturally, all of this requires road access, but without,
these benefits to citizens and wildlife would not be possible.

Westvaco: Wildlife Management

Hopping over the Southwest, the Lake States, and the Northeast, we find one more case history of unexcelled forestland stewardship. Westvaco, headquartered in New York City, quietly and effectively goes about the job of managing renewable forestland ecosystems, another fact that is not picked up by the media.

For years Westvaco has implemented a partnership with
nature. In its view forestland ecosystems, farms, and wildlife go
together "naturally."[36] This was exemplified with the signing in June
1991 of the Memorandum of Understanding which established the
Westvaco Wildlife Management Area on the banks of the Mississippi River near Wickliffe, Kentucky.

Active participants in this wildlife management program
include the Kentucky Department of Fish and Wildlife Resources
and Ducks Unlimited. The program is one element in a joint effort
of the United States, Canada, and Mexico which together sponsor
the North American Waterfowl Management Plan. Our coordinat-
ing agency for this effort is the U.S. Fish and Wildlife Service.

Westvaco's contribution to the program includes the utilization
of over three thousand acres of prime forestland ecosystems. This
land will stay on the tax rolls and will continue to supply raw material for the pulp and paper complex in Wickliffe, but some of the land
will be planted in corn, milo, soybeans, and other farm crops to provide food and cover for a multitude of birds and animals.

Portions of the Westvaco Wildlife Management Area, which

lies along the Mississippi Flyway, will be closed between October 15 and March 15 each year to allow migratory waterfowl to rest and feed undisturbed. Except for this closure, hunting and fishing are permitted. Initial projects to accommodate wildlife, such as drilling for water, the placement of water control structures, and the creation of additional food plots have been completed.

This marriage of environmental and economic benefits in the Wickliffe area is just another in a long line of cooperative programs that Westvaco has undertaken with groups such as Trout Unlimited, the Nature Conservancy, the Ruffed Grouse Society, and the National Wild Turkey Federation as well as various federal, state, and local agencies.

At the dedication, a senior member of the Westvaco management team noted, "Westvaco is committed to a multiple-use forestry program including wildlife, recreation, and watershed protection" as well as supplying raw material to various production facilities from its 1.5 million acres of forestland.[37]

Westvaco also helps more than 2,700 nonindustrial forestland owners. Inaugurated in the 1950s, this private landowner assistance program, one of the first of its kind, is appropriately called Cooperative Forest Management. With constant professional assistance provided in the field at no charge by Westvaco, these smaller landowners now care for more than 1.3 million acres of renewable forestland ecosystems. Once again, this is ecosystem management at its best.

Fittingly, Westvaco was the first corporation to receive the National Wetlands Conservation Award from the U.S. Department of Interior, Fish and Wildlife Service in a June 1992 ceremony conducted in Washington, D.C. Westvaco was honored in this case for its work with the Kentucky Department of Fish and Wildlife Resources in establishing the Westvaco Wildlife Management Area on the Mississippi River.

Those interested in the work of Westvaco wildlife biologists and natural resource professionals are welcome to visit. The Timberlands Division, Central Region office in Wickliffe offers a "Wildlife & Westvaco at Wickliffe" folder, a handsome series of eye-catching photographs and informative details for nature-lovers.

• (Photo 37): This photo in black and white does not begin to show the detail in the large color rendition in the folder just mentioned. The white specks are egrets and herons enjoying themselves on a small portion of this newly established Wildlife Management Area. As you can see, the waterway has a buffer zone of natural growth between farmlands on the right and plantations of cottonwood and sycamore on the left.

IS A POSITIVE VISION OF FORESTLAND STEWARDSHIP POSSIBLE?

Setting aside our renewable forestland ecosystems in designated reserves is not stewardship. Rather, stewardship is sharing our concern for trees and forestlands in a positive way. In my view, ecosystem management offers a golden opportunity to do this.

This new concept is golden because it will alert average citizens and professional resource managers to the fact that trees and forestlands are indeed ecosystems, encompassing the air, the water, what is on the ground, and what is under the ground.

Remarkable progress in this direction has been made since World War II. Professional foresters working closely with silviculture and soil scientists, and in more recent years with aquatic and wildlife biologists, hydrologists, and other scientists, have for years concerned themselves with ecosystem management, though they didn't call it that.

This should come as no surprise. Since midcentury, billions of dollars have been invested in renewing our forestland ecosystems, both in the public and private sector. Yet unfortunately—tragically—stewardship of our public forestlands has become all but hopeless. As long as the law forbids the conservation and management of renewable forestland ecosystems in the public sector, stewardship will be impossible.

Concerning the priorities in the private sector, and what drives them, the Forest Products Industry's first priority has been to meet the demands of society all the way to the final consumer. Nearly as high a priority is nurturing a resource base that is solar driven and renewable. Obviously, huge investments in the resource base have not been made at random, and long before the word "ecosystem" became fashionable, professional forestry included the care of soils and microscopic mycorrhiza along with a wide range of plant life. Site indicators—that is, plant life which indexes the richness of the soil—have been important from day one.

Maintaining the health of ecosystems to control insects and disease, reducing fuel loads to prevent wildfire, and maintaining healthy watersheds are essential steps in stewardship. In order to do these things, access is absolutely necessary.

Of course, healthy ecosystems are a bonanza for recreation and wildlife. And for communing with nature and attaining a spiritual uplift found only in nature, a healthy natural environment far surpasses one full of dying, dead, and down trees so characteristic of the forestlands of our National Park System, our Wilderness System, and other set-asides. Neglected ecosystems are not only depressing, they are dangerous.

A positive vision of forestland stewardship is possible. But when our National Forest System is burdened with 226 *principal* laws that often conflict, it is not an easy task. Nonetheless, common sense and the implementation of ecosystem management should help achieve this goal.

Access will be greatly expanded with ecosystem management because care will be mandated for all our forestlands, including our National Park System, our Wilderness System, and other set-asides.

To wrap up: There *is* a difference between the enabling legislation that evolved during the first fifty years of forestland conservation and management, and the gridlock restrictive legislation that has sprung up since World War II. In consequence, the supply of raw material from commercial forestlands in our National Forest System is close to zero. Congress is totally disregarding the long-standing charter of the Forest Service that promised to supply our raw material needs.

Since midcentury, we have been successfully renewing our forestland ecosystems, after having exploited them for over three hundred years with the cut-and-run method of the Forest Products Industry and the clearing-and-burning system of agriculture.

But despite overwhelming evidence to this effect, antiforestry advocates have convinced the general public that the Forest Products Industry is destroying our forestland ecosystems in nineteenth-century fashion. Evidently anything is possible when dealing with perceptions rather than reality.

It is now time to deal with reality, to discern what *is* and what *is not* stewardship, and to get on with the job of implementing ecosystem management.

PART II

FORESTLAND ECOSYSTEMS AND OUR FUTURE

4

NEGATIVE AND POSITIVE
CASE HISTORIES

IT HAS BECOME quite apparent that the good news of renewing our forestland ecosystems is not considered news. Instead the emphasis is on confrontation and bad news. That, it seems, is what sells. But in order to promote positive action on behalf of trees and forestlands, it will be necessary to seek the truth. And truth is illusive.

FORESTRY: PARTING THE CURTAIN OF PERCEPTIONS

Professional forestry has been damaged by an overdose of misleading rhetoric. Consequently the resilient and dynamic characteristics of our renewable resources are little understood by the general public. Bombarded by the propaganda of antiforestry advocates, the general public has overlooked the single most important positive trend in forestland conservation and management in this century— the renewal of our forestlands rather than their elimination. On the contrary, by spreading clever but false perceptions, antiforestry advocates have convinced the general public that industry is destroying forestland ecosystems.

We must look beyond perceptions and become better informed about the special interests that foster them. Let us look at just two examples of why this is necessary.

Clearcutting: a Negative Perception

For the temperate marine forestlands along the Pacific Ocean and for much of the Southern pine region, current science and technology indicate that the widely used practice of clearcutting is the best prescription for renewing forestland ecosystems.

Of course, there are exceptions. One of the best examples of pushing clearcutting beyond its usefulness can be found in the Bitterroot Mountains of Montana. Generally speaking, in open pine country on the dry side of the mountains and on the hot, southern and southwestern exposures, clearcutting should be avoided. But on the wet side of the mountains, there is no better method of bringing back forestland ecosystems to the way they were before harvesting than clearcutting.

The advantages of clearcutting are portrayed clearly in textbooks, yet antiforestry activists continue to preach the negative aspects, just as they have for many years.

To compare a clearcut harvest in the Far West with clearing tropical forests in the Amazon Basin or with deforestation in any sense of the word, is a deception of the first order. And to push legislative initiatives at the federal and state level outlawing clearcutting is an even worse infraction. (On 12 May 1993, a ruling by federal Judge Robert Parker ended clearcutting on four National Forests in Texas.)

For a shot of reality, I recommend David M. Smith's textbook, *The Practice of Silviculture*.[1] Smith, the Morris K. Jesup Professor Emeritus at Yale University, is the 1990 recipient of the American Forestry Association Distinguished Service Award. In the association's words, Smith is the "leading silviculturist in the United States and perhaps the world. His understanding of the ecology and silviculture of mixed-age and mixed-species forests and his ability to make these clear to others is unsurpassed. *The Practice of Silviculture* is the standard text and reference in North America and is used throughout the world."

Smith dedicated the most recent edition, the eighth, "to Ralph C. Hawley, pioneer American silviculturist," author of the first five editions, dating back to March 1921. Smith remembers: "It was Hawley's view that silvicultural practices should be prescribed by foresters who had their feet on the ground, literally and figuratively, with their minds at work."[3]

Smith said he had "attempted to carry Professor Hawley's philosophy into the present. Even more emphasis has been put on applying scientific analytical reasoning to the formulation of silvicultural solutions to the diversity of management problems that foresters face. These problems are still construed as including both biological and socioeconomic considerations."

In the text, Professor Smith makes a strong case for the "justifiable application of clearcutting," and then goes into considerable detail. For example:

> Much of the clearcutting and planting being done in the western United States involves stands that are too old and decrepit to endure regimes of partial cutting. . . . The exigencies of timber harvesting are most likely to favor clearcutting and planting if the machinery is ponderous, roads are costly, and the terrain is difficult.

In the section "Evolution of Silviculture in the West Coast Douglas-fir Region," he addresses the selective cutting experiments in the 1930s:

> These circumstances set the stage for an episode of ill-fated efforts to prolong the lives of old-growth stands by light partial cuttings referred to as "selective logging." The cuttings were intended to operate as selection cuttings to create uneven-aged stands. . . . By the 1950s the whole episode was recognized as a fiasco to be quietly forgotten.

Two photos depict "before" and "after" selective logging of old growth on Washington's Olympic Peninsula in the mid-1930s. In the caption, Smith comments, "Unfortunately this alternative to clearcutting started the accelerated decline of most stands of this kind."

In the chapter "Guilt by Association," Smith states:

It is a common belief that tree cutting caused the sad history of soil damage and erosion in the Mediterranean Region, East Asia, and many wet and dry tropical areas. In virtually all such cases the damage to the soil resulted from what people did *after* removing the forests [emphasis his].

Smith concludes: "Goats have probably destroyed more forests and caused more erosion than anything else." He cautions that it makes no sense to be complacent about the real culprits while "decrying actions taken to grow trees, which are the best soil defenders available."

Smith underscores this point with the statement: "The view that clearcutting causes physical erosion of the soil is almost groundless."

In concluding his discussion of clearcutting, Smith addresses the entire forestland ecosystem in "Defense by Revegetation" by emphasizing:

The timber-production forest managed for sustained yield with all age classes from one year to that of rotation age would come close to the perfect system for keeping soil nutrients, nitrogen, and carbon cycling within the whole forest.

Moreover, "There is no better means of conserving the renewability and productivity of terrestrial ecosystems than keeping the growing space full of vigorous forest vegetation." And finally, "Since successful clearcutting is that which results in prompt revegetation it is not necessarily more harmful to the environment than other methods."

When Smith speaks of a forest rotation age which would come close to the "perfect system," you can visualize a one hundred acre forest with a one hundred year rotation age and a harvesting prescription calling for one acre to be harvested each year. You can now picture that ninety-nine acres will always have vigorous growth, from seedlings to a mature ninety-nine-year-old stand of old growth. Even the one acre harvested each year would have "vigorous" forestland vegetation.

Of course, what you can actually see, particularly in the Far West, are examples of clearcut stewardship which are far from the perfect stewardship alluded to in Smith's textbook. When stewardship is ignored, I side with the critics. But current legislative mandates banning all clearcuts on public lands is at best counterproductive. It is a clear case of "throwing out the baby with the bath water."

If clearcutting had been banned after World War II, the supply miracle would never have occurred, the renewal of our forestland ecosystems would not have been possible, and regulatory paralysis would have surfaced long since. It is vital to the interests of society at large that it recognize that clearcutting can and should be an integral part of the "perfect system" of stewardship on public as well as private forestlands.

Wetlands: How Will We Get at the Truth?

As with the clearcutting controversy, the issue of wetlands has spawned a great deal of misleading rhetoric.

In colonial times there were some 226 million acres of wetlands, according to the U.S. Fish and Wildlife Service. By the 1980s, original wetlands covered no more than an estimated 105 million acres. Then in 1989 one of the most extraordinary events in conservation history took place. Several government agencies increased the nation's wetlands by adding millions of acres of "dry wetlands" which became known as "jurisdictional wetlands."[4] The outcome, of course, is that the wetlands issue has become a choice "political football." And a confusing issue it is. What, for example, constitutes a wetland?

When the issue of redefining wetlands was announced on network news in August 1991, viewers were served a large dose of hype. We saw lovely panoramic views of ponds, lakes, lagoons, and flooded marshlands which were not involved in any way in the controversy. Commentators suggested that these bona fide wetlands were at risk, when nothing could have been farther from the truth.

According to the current regulatory definition, however, just about every Little League field and golf course in the United States can be classified as wetlands and subject to federal jurisdictions that are onerous in the extreme.

Here is one example. In 1988, Ocie Mills of Navarre, Florida, spread sand on a one-acre homesite, "a damp indentation in the ground," with the state's permission. "Whereupon the U.S. Army Corps of Engineers arrested Mills and his son, Carey, for filling-in wetlands. They served 19 months of a 21-month sentence in federal prison."[5] Mills has appealed his conviction.

In 1993, U.S. District Court Judge Roger Vinson issued a twenty-one page decision in the case. "Although he did not overturn the conviction, he did state that the Clean Water Act has been twisted into wording 'worthy of Alice in Wonderland.' He indicated that the land Mills was trying to build a home upon was clearly *not* a wetland." He allowed that "Congress does have the authority to delegate the definition of 'navigable waters' to the Corps, even if the definition is absurd."[6] Further, according to an 8 April 1993 *Wall Street Journal* article, Judge Vinson "warned that it is dangerous for Congress to abdicate its power to define a felony to an un-elected administrative agency."

The focus of the confusion is on the technical definition of "jurisdictional wetlands." There are two federal agencies responsible for wetlands regulations: the Environmental Protection Agency and the Army Corps of Engineers.[7] Their original wetlands definition was, "those areas that are inundated or saturated by surface or ground water at a frequency and duration sufficient to support, and that under normal circumstances do support, a prevalence of vegetation typically adapted for life in saturated soil conditions. Wetlands generally include swamps, marshes, bogs, and similar areas."

This is an accurate definition. Federal agency representatives, however, met in 1989 behind closed doors, without public input or review, and further defined wetlands using three criteria; hydrology, soils, and vegetation.[8] The hydrology criterion says that if land is saturated for seven consecutive days *within* 18 inches of the surface it is wetland. Note, it is not "wet" wetland as pictured in the media. That is, it is not necessary for splashable water to be on the surface. In fact, it is not necessary for there to be any water at all. Rather, it is only necessary for the soil to be saturated within 18 inches of the surface.

This definition expanded by many millions of acres what was thought to be wetlands in the 1980s. And it provided another grand opportunity for fund raising. Environmental activists, with the support of important segments of the media, are now intent on maintaining the definition of dry "jurisdictional wetlands."

The recently proposed revisions to the clandestine 1989 definition of wetlands do nothing more than attempt to let farmers continue to farm and foresters continue to manage forestlands. In addition, developers would not be held to federal regulations for wetlands when in fact the land in question is not wet.

The proposed revisions would apply a small dose of common sense to this controversial issue and withdraw some obviously dry land from the overreaching federal definition of "jurisdiction wetlands."

In fact, the proposed changes do not change dry wetlands that much. Among other things, they call for *inundation* of the land with water for fifteen days (a wet wetland) or *saturation to the surface* for twenty-one days (a wetland without water). Under either definition, waterfowl would have a rough and dry landing nearly every day of the year on "jurisdictional wetlands."

The most troubling aspect of this confusing issue is that many forestry operators in Florida, Georgia, and North and South Carolina are finding that upwards of 90 percent of their forestlands are "jurisdictional wetlands."[9]

And though forestry is supposedly exempt under Section 404 of the Clean Water Act, operators are now faced with a host of aggressive government agency representatives looking over their shoulders and working hard to narrow the practices that are exempt. Considering the success of antiforestry advocates in setting aside commercial working forestlands in designated reserves, professional foresters are understandably concerned about the 1989 clandestine wetlands definition.

URBAN FORESTRY: POSITIVE CASE HISTORY

Thanks to the American Forestry Association, we already possess a plan for urban forestry which takes us into the future with real promise. Urban Forest Councils are now established in every state, the District of Columbia, and Puerto Rico.

This effort is outlined in the December/January 1992 *Urban Forests*, a publication of the association.[10] In an editorial, Gary Moll, vice president for Urban Forestry, notes that more councils, thirty in number, were established in 1991 than during the entire 1980s. In November 1991, members of the National Urban Forest Council gathered in Los Angeles for the Fifth Urban Forest Conference. There they adopted five new policy objectives:

- Greatly increased public recognition of trees and related resources as integral and important parts of the nation's urban infrastructure, which must be included in federal, state, and local government and private construction funding and maintenance funding for such projects as highway building and rebuilding, public housing, community improvement, and private development.

- Substantially improved understanding by legislators and others who influence public policy about what urban and community forests are and that investments in our nation's urban and community forests can provide cost effective benefits. The investment costs, nationwide, will involve billions of public dollars.

- Increased federal funding to support urban and community forestry efforts.

- Increased funding for urban forestry research, with emphasis on quantifying tangible benefits of urban forests, especially as they relate to energy conservation and water quality and conservation.

- Formation of active, broadly based, local urban and community forest councils on a nationwide basis.

This issue of *Urban Forests* is full of helpful hints. For example, "Success in gathering a group of people. . . does not guarantee that everyone is going to agree. Each person brings something different to the council, that's why it works."

Another: "The power of an Urban Forest Council relies on the development of ideas 'whose time has come,' and the promotion of those ideas among the government officials and members of the public who have the ability to influence national, regional, state, and local policies."

Some other suggestions for operating a successful Urban Forest Council are:

- Set primary goals and objectives that are limited in scope.

- Select goals and objectives that are achievable and can be monitored.

- Establish committees or task forces.

- Brainstorm ideas that will educate your communities about the importance and value of trees and forests.

- Invite legislators to your meetings.

- Write letters to decision makers advocating your council's positions.

- Produce posters, brochures, and other educational materials.

- Don't get stuck on procedural matters or five-year plans.

- Quickly get past administrative issues and get down to building community alliances and programs that encourage laypeople to support urban forests and plant and care for trees.

In this same issue, Donald C. Willeke, president of the American Forestry Association and chairman of the National Urban Forest Council, shares some thoughts. Willeke has been involved for many years and has no doubt addressed more groups of concerned citizens at the grass roots level than any other individual. He says that the most frequently asked question by urban forest enthusiasts is, "How can we get authorization for such local councils?" Willeke's answer: "If you can get your city council to authorize an urban forest council, or if you can get your mayor or city manager to appoint one even without city council authorization, that's fine. But if you can't get either quickly, then *just do it!*"

Willeke further notes, if you "wait for 'proper authorization' it may never come." And finally he assures, "A council grown from the bottom up rather than authorized from the top down is usually composed of more dedicated people, and is likely to be more representative of the many constituencies that are interested in a community's trees."

The good news for urban forestry is that our oldest citizens' conservation organization has been able to identify problems, analyze solutions, and stimulate urgently needed action at the grass roots level. In short, with urban forestry we are sharing concern for trees and forestlands in a very positive way.

Though much progress has been made, considering the immense scope of the task at hand, we are still only scratching the surface. Even with all the encouraging progress in urban forestry, as with industrial and nonindustrial forestry, it is imperative that we gain the attention, and, we hope, the enthusiastic participation of the general public.

DRUCKER ESSAY: A POSITIVE OUTLOOK

Growing trees requires the support of all citizens. Citizen involvement was the subject of a succint essay written by Peter Drucker in the *Wall Street Journal*.[11]

Drucker points out that "virtually every success we have scored has been achieved by nonprofits," as, for example, the American Heart Association, the American Mental Health Association, Alcoholics Anonymous, the Salvation Army, parochial schools, and so on. And Drucker observes, "The nonprofits spend far less for results than governments spend for failures." Nonprofits have "the potential to become America's social sector—equal in importance to the

public sector of government and the private sector of business. . .
.The delivery system is already in place."

Drucker notes that currently there are some 900,000 nonprofits—thirty thousand added in 1990 alone—which dedicate themselves to "local action on one problem" such as "tutoring minority children, furnishing ombudsmen for patients in the local hospital, helping immigrants through government red tape. . . .There are now some 90 million Americans—one out of every two adults—working as 'volunteers' in nonprofits for three hours a week on average; the nonprofits have become America's largest 'employer.'"

One of Drucker's most telling points is that volunteers do not look upon their work as charity. Rather:

> They see it as a parallel career to their paid jobs and insist on being trained, on being held accountable for results and performance, and on career opportunities for advancement to professional and managerial— though still unpaid—positions in the nonprofit.
>
> Above all, they see in volunteer work access to achievement, to effectiveness, to self-fulfillment, indeed to meaningful citizenship. And for this reason there is more demand for well-structured volunteer jobs than there are positions to fill. . . .

Further, Drucker predicts that "within 10 years, two-thirds of American adults—120 million—will want to work as nonprofit volunteers for five hours a week each, which would mean a doubling of the man- and woman-power available for nonprofit work."

Drucker contends that present IRS initiatives are stifling and calls for a change in the government's attitude, with tax incentives to increase the effectiveness of nonprofits. "The real motivation for such [stifling] actions is the bureaucracy's hostility to the nonprofits—not too different from the bureaucracy's hostility to markets and private enterprise in the former Communist countries.

"The success of the nonprofits undermines the bureaucracy's power and denies its ideology. Worse, the bureaucracy cannot admit that the nonprofits succeed where governments fail." Drucker's analysis goes to the heart of why perceptions are so much more important than reality inside the Beltway.

But let's think for a moment beyond parting the curtain of perceptions. With 120 million adult volunteers, giant steps can be taken by urban forestry professionals and volunteers working hand in hand in every state to nurture trees that are in urgent need of care. At the same time volunteers can direct their attention to the urgent need for ecosystem management in our rural forestlands.

The concept of ecosystem management is timely. With an enlightened understanding of what *is* and what *is not* stewardship, we can give new meaning and force to nurturing our renewable forest-land ecosystems. Finding common ground and breaking through regulatory gridlock can only come about through the concerted efforts of citizen conservationists, members of the media, legislators, and leaders of the Forest Products Industry.

5

TREES AND CORPORATIONS

PRIMARY PRODUCERS IN the Forest Products Industry—whether in the pulp and paper sector or the wood products sector—have an identity problem. (In January 1993 the two sectors merged into one national organization—the American Forest & Paper Association.) It almost seems as though they don't understand that they are members of the Forest Products Industry . For decades, this lack of understanding on the part of industry's CEOs has played directly into the hands of antiforestry advocates who drum the idea of the "timber industry" into the minds of the general public.

"TIMBER" INDUSTRY OR FOREST PRODUCTS INDUSTRY?

If you should ask the question as I have in the Far West, "Do you know very much about the Forest Products Industry?" often the answer is, "No, I do not." And then, "Do you know much about the 'timber' industry?" Most often the answer is, "Oh, yes, we are surrounded by 'timber' companies." In short, very few people know anything about the Forest Products Industry.

This of course includes the majority of people in the news media, not to mention financial analysts, government employees, scientists, educators, students, members of trade, technical and professional groups, members of the environmental community—and employees of the Forest Products Industry at every level.

Obviously, the Forest Products Industry sorely needs recognition of its accomplishments and affairs, and to teach of these in a positive way. But today, even its existence is unrecognized.

Two of our most widely read newspapers, the *Wall Street Journal* and the *New York Times*, have a daily index of companies featured in each edition. *Barrons*, the business and financial weekly, does the same, as does *Fortune* and other publications. But Forest Products Industry firms are rarely mentioned.

This neglect pertains although since 1980, in the pulp and paper sector of the Forest Products Industry alone, over $110 billion has been invested in new facilities.[1] These capital investments are for hardware, processing technology, quality control, productivity, and compliance with a variety of environmental standards and regulations.

And this expenditure is typical of the entire Forest Products Industry over the past fifty years. Each day, citizens everywhere reap the benefits from these investments, but the CEOs of this world-class industry have neglected to inform the general public where and how these benefits come about.

World-class indeed! Solar-driven, renewable raw material base, provider of thousands of our essential daily needs. Yet the index to the ninth annual *Fortune* "Investor's Guide" (October 1993) lists 220 companies and not a single Forest Products Industry company makes the grade![2] In truth, recognition is almost nil.

But recognition aside, the Forest Products Industry is important for other reasons as well. There is already sufficient data indicating that forestland ecosystems, i.e., biodiversity and critical habitat, are being "protected" only by the common sense management of the private sector. Yet the federal government increasingly encroaches on the industry, tightening central controls, which in turn dissipate ecosystem values—not unlike what happened in Eastern Europe.

Antiforestry advocates nevertheless continue to assert that the "timber" industry's only concerns are commercial and that noncommercial values can be safeguarded solely by federal controls. Thus we see regulatory gridlock tighten while industry leaders make $110 billion investment decisions and assume while doing so that raw material for primary mills will be readily available.

"TIMBER": WHAT DOES THE WORD MEAN?

The word "timber" is so overused, it now means nothing. Citizens do not consume "timber." Rather, "timber" is the commercial word used best to describe raw material. "Timber" means forestlands, stumpage, trees, logs, and the like.

It also precedes words such as profits, resources, country, company, sales, supply, harvesting, wolf, baron, beast, and so on. Writers for the media and for fund-raising solicitations lean heavily on the word "timber" to hide their lack of knowledge of forestry, fully understanding the negative connotations of the word. Citizens enjoy forestlands, watersheds are protected by forests, critical habitats for animals and plants are found in forestlands, and owls live in forests—not "timber." All bad things turn on the word "timber." All good things derive from the word "forest."

The worst abusers of the word "timber" are the writers of antiforestry fund-raising letters. The word has contributed greatly to contriving the false issue of harvesting. It has also helped build the case for the "wanton destruction" of forestlands, critical habitat, recreation opportunities, and nature itself. As a result, anyone having anything to do with the "timber" industry is beneath contempt.

Amazingly, the next worst abusers of the word "timber" are members of the Forest Service family. Overuse of the word in Forest Service literature has become a bad if unconscious habit. For example, in the Glossary of Forest Inventory Terms outlined in the 1989 *USDA Forest Service RPA Assessment*, "timber" is used repeatedly as a crutch word.[3]

Forest Products Industry people are also careless with their use, and thereby increase their vulnerability to misleading rhetoric. A typical example is found in the January/February 1992 National Forest Products Association bulletin, *NFPA in Focus*. At that time, the NFPA was the industry's flagship organization in the nation's capital.

The feature article, "Owl Habitat Designation Hits Hard," included terms such as "timber workers," "timber industry," "timber communities," "timber-dependent counties," "timber sales," and so forth. In eight short paragraphs, we are "timbered" to death.

When articles such as this refer to "timber communities," they seem to imply that people in small towns are only interested in the "timber" that is to be cut, and forestlands be damned. But in reality, people in small communities play a key role in renewing our forestlands.

Trade journals and even scientific and professional journals add to the confusion. The copy in trade journals is strikingly poor. When in doubt, they slip in the word "timber." If the word were removed from the industry lexicon, some trade journals would be forced to close down.

Antiforestry advocates go one step further. They not only paint a picture in which those who live in "timber communities" are the destroyers, but magically transform "timber" into "forest" when speaking of a home for wildlife, plant life, and other noncommercial values, complete with such adjectives as pristine, ancient, crown jewels, and national treasure.

COMMUNICATING THE INDUSTRY'S POSITION

It is not only possible but essential to rectify the situation by persuading the general public to think of the industry not as the "timber" industry, but as the Forest Products Industry.

There is in place an infrastructure to move the Forest Products Industry smartly into the twenty-first century. The recent consolidation of the pulp, paper, and wood products sectors is expected to enhance progress greatly as it has paved the way for industry leaders to fund a landmark communications effort.

Yet the industry seems intent on carrying on with business as usual from its base in the nation's capital. If so, the industry will continue to have grave difficulties communicating its position.

When individuals, companies, and industries flock to the nation's capital, they appear to endorse the process of increasing central controls and regulatory gridlock. The challenge for the industry is to craft communication goals and programs that will impart to the general public what is actually taking place in forestland conservation and management. But in order to do so, industry leaders must make a bold move. They must move their communications effort away from the capital and away from a culture that espouses perceptions in place of reality.

FOREST PRODUCTS INDUSTRY INFORMATION CENTER

As part of this bold move, I should like to recommend the establishment of a Forest Products Industry Information Center. The Center will sharply focus the industry's task of educating the general public about the positive contribution that forestland conservation and management make to our environment. The Center will come to grips with reality and concentrate on the task of how best industry can serve society at large with the countless benefits of our renewable forestland ecosystems.

Center Location, Equipment, and Staff

The Center will be located in mid-America and will divorce the industry from the Beltway and urban America.

The Center will be equipped for processing information on a global scale.

As for the staff, young people will be recruited by offering them an opportunity to make a major positive environmental impact in all areas related to trees and forestlands. Funds should not be spared to attract young people who are bright, highly motivated, articulate, conscientious, dedicated, and energetic. Funds should be available to attract the foremost communicators worldwide.

Center Priorities

With a few notable exceptions, industry communication programs have tended to be spotty, inconsistent, and ineffective. Industry leaders have left the environmental high ground to antiforestry advocates, a tactical blunder. Funding the landmark communications effort and the Information Center would be the first significant step to right this situation by communicating the industry's position in a consistent and effective manner.

But the number one priority would be to announce to the general public, here and abroad, that the Forest Products Industry in the United States exists. The fact that it exists, and the reality of its positive impact on the daily lives of citizens will surprise many within the industry as well as those outside.

Center Priority: What Is the Industry Universe?

The antiforestry advocates' position on the environmental high ground, it seems, is unassailable. Supported by important elements of the media, these advocacy leaders have convinced the general public that they are the protectors of our environment whereas the Forest Products Industry "destroys" it.

Antiforestry advocacy leaders, moreover, have been single-minded in their narrowly focused agenda to discredit the industry, and industry leaders appear unable to cope; as a result, their credibility of the past fifty years, or 350 years, has been severely damaged.

As matters now stand too many people within the industry do not realize that they are an integral part of it. Little thought is given to the industry universe. How, for example, does the individual fit in? How does a company, large or small, fit in? What is the true role of the Forest Products Industry in society?

Genevieve Capowski, in a lead article, "Where Are the Leaders of Tomorrow?" in *Management Review* (March 1994), outlines an interesting picture of leaders vs. managers. Leaders, she says, work from the heart and soul; they are visionary, passionate, creative, flexible, inspiring, courageous, experimental, and independent. Managers, on the other hand, work from the mind; they are rational, persistent, analytical, structured, authoritative, and stabilizing.

This description of managers fits Forest Products Industry leaders who over the past fifty years have not only produced a miracle, but shifted from the practice of exploiting our forestlands to renewing them. The Information Center concept challenges industry leaders to adopt a visionary approach in order to solve the incredibly difficult problem of regulatory gridlock. The Center will provide a timely means of doing this and will recognize the importance of the role of all participants within the Forest Products Industry universe.

The visionary approach for the twenty-first century must be based on the reality that the only reason for the existence of the Forest Products Industry is to serve society. Far too many industry people are unaware of their role in the universe and how their work contributes to the well-being of all citizens.

Peter Drucker alludes to this fundamental in *Practice of Management.* "If we want to know what a business is," Drucker said, "we have to start with its *purpose.* And its purpose must lie outside of the business itself. In fact, it must lie in society since a business enterprise is an organ of society." [4]

My experience suggests that one of the basic flaws of the forest products business is the overlapping vested interests in the many "sandboxes" scattered throughout the industry. Let me try to illustrate this problem.

Say you are the CEO of a leading Forest Products Industry company. You look out the window of your office and realize you are surrounded by concentric circles of sandboxes. There are people with individual vested interests in each box. Dozens and dozens of sandboxes. Placards in some of the sandboxes read: "Stay in your own box." The leaders in the sandboxes have a pact that goes something like this: "I won't pump gas if you won't sell newspapers." In other words, don't mess up the play in my sandbox.

With so many players and so many sandboxes, where do they all fit into the universe? Creditable solutions to individual firms and to industry problems are illusive. At the CEO and director level, consulting fees in seven figures are an accepted part of doing business. Participating in various business round tables is routine, as is hiring public relations firms and advertising agencies. But despite all the money spent, the problem remains—no one really grasps the full picture of the industry universe.

Consulting firms have become experts on what works and what doesn't work by observing and listening to all the people throughout the universe. When an assignment is completed, no one knows as much about the business and its place within the universe as the outside consultant. CEOs and directors are astounded by the brilliance of their observations and recommendations.

Workers in the mills and forestlands are also disadvantaged when they lack a feeling or a sense of belonging to the greater universe. Loggers, for example, are essential. Vigorous, healthy forestlands cannot be renewed unless they are harvested. And such neglected forestlands as those in Idaho, Montana, and Eastern Oregon can only be made well by the expertise and dedication of loggers.

Fallers, the rigging crew, haulers, and yes, road builders, as well as those who prepare the soil for tree planting, and those who do the planting are all essential team players. Without them it is not possible to renew our forestland ecosystems. At the moment, far too many of these important men and women are unsung heroes. That will no longer be the case when the Center focuses on the real story of growing trees.

Part of the problem of fitting the function of the Forest Products Industry into its place in the universe or its reason for being is that the words of the Peter Druckers and the Tom Peters of the world are too often ignored. That is, every business should strive for the ideal commitment to "best customer service."

Therefore, all meetings, programs, travel, and expenditure of funds not supportive of or directly related to best customer service should be curtailed. This outlook fits into the theme of the Information Center—to relate the purpose of the Forest Products Industry to societal needs and to assist all industry operators to learn how best to focus on their reason for being—to serve society.

It is essential that forestland ecosystem management be carried out in a way that will meet all the needs of society at large.

And it is essential that manufacturing and marketing functions be showcases for how the industry emphasizes concerns for quality, value, safety, service, environmental controls, and other attributes that citizens can view with pride.

Center Priority: Forestland Inventory

The Center would address a host of housekeeping items which have been left untended, largely because the industry has been preoccupied with working the supply miracle. The Center staff will place forestlands in a perspective that can be easily understood by the general public.

This will require an accurate, complete, and up-to-date inventory of our forestlands. With the available technology, establishing this data base should not present a problem. Center planners would coordinate the work which would be contracted out.

Center Priority: Player Roster

This all-important industry, strangely enough, does not have a data base which lists all of the people directly involved. Major association executives in the wood products sector have good information on the 20 percent who do 80 percent of the business, but beyond that they know far too little.

The Information Center must rectify this situation for communication purposes. The 80 percent who do the remaining 20 percent of the business are as important as the major producers. Industry CEOs have had trouble understanding this. They tend to overlook the role of small, family-owned operations.

The Center must also be able to communicate with decision makers in every company in the Forest Products Industry, regardless of place and size. Again, establishing this data base with present technology should not present a problem. Center planners would coordinate the work which would be contracted out.

Without an up-to-date, accurate forestland inventory and a comprehensive data base of operators immediately dependent upon the inventory, the industry is ill prepared to enter the twenty-first century.

Once this critical information is in hand, Center staff could search out those businesses indirectly dependent on the forestland inventory; this data base would include secondary producers and the many other businesses which would be disrupted should raw material from our forestlands no longer be available.

With these tasks completed, the Information Center staff will be able to address those who form the link between our forestlands and consumers of forest products. Citizen conservationists in turn will have a source of information concerning the truth about forestland conservation and management.

For example, these conservationists will have the opportunity to learn more about how management of forestlands enhances ecosystems and biodiversity in contrast to allowing nature to take its course. However, Center staff will be active and not wait for inquiries. The daily accomplishments in our forestlands, mill operations, and distribution networks will be shared far and wide. In due time, the media will catch on to this positive approach.

Center Budget

We have looked at the concept of an Information Center. But what will it cost and who will pay the bill? In 1992 the consolation of the pulp, paper, and wood products sectors required a bold move by industry leaders. It is now possible for the industry to speak with one voice as the landmark communications effort is implemented. To further this effort the CEOs of the major corporations in the Forest Products Industry must bite the bullet one more time.

The seed money required for the Center will be in tens of millions of dollars, and its operations will continue to cost in the range of tens of millions annually. Projecting annual budgets for the Center into the next century, CEOs will be compelled to think in terms of hundreds of millions of dollars. Where will the money come from? Let me try to place the cost of the center in perspective. In our consulting work, we monitor Forest Products Industry companies by annual sales as follows:

Table 6
FOREST PRODUCTS INDUSTRY COMPANIES

Grouped by Sales Volume

Group I	Over $2 billion
Group II	$750 million to $2 billion
Group III	$100 to $750 million
Group IV	$25 to $100 million
Group V	$1 to $25 million
Group VI	Under $1 million

Descriptive terms as well as numbers would be appropriate for these groups, i.e., Giant, Very Large, Large, Medium, Small, and Very Small. The implications of size are not readily apparent. A family business in Group V (Small) generating annual sales volume of $15 million may not be considered small in its own neighborhood, but compared to other operations, it is small. Relate the size of a business to our sandbox analogy. Across the country within the Forest Products Industry, we have situations where small businesses have banded together by the hundreds in a sandbox to protect themselves and their vested interests from large operators in other sandboxes. This pattern of "them" and "us" with regard to size is not helpful from an industry standpoint.

The bulk of the money for the Center budget will come from Groups I, II, and III; not from Groups IV, V, and VI. For the purpose of this illustration, we will talk about only thirty companies. Presently, we monitor sales and net after-tax profits of these publicly held firms as their figures appear quarterly in the *Wall Street Journal*. Of these thirty companies on our list, fourteen are in Group I, eight in Group II, and eight in Group III.

Sales and earnings for these companies are listed in Table 10 in the appendix. In 1991, total sales were $91.3 billion, with net profits of $1.3 billion or 1.4 percent of sales. On the basis of earnings in 1991, we cannot expect funding from these companies for the Information Center.

A five-year outline of sales and earnings provides a more representative illustration of the source for Center funds. This five-year recap by company is outlined in Table 11 in the appendix. A summary of sales and profits for these thirty companies follows:

Table 7

TOTAL ANNUAL SALES AND NET AFTER-TAX PROFITS FOR 30 FOREST PRODUCTS INDUSTRY COMPANIES 1987 TO 1991 ($000,000)

Year	Sales	Net after-tax Profit
1987	$72,733	$4,397
1988	81,835	6,459
1989	90,681	6,387
1990	94,681	4,107
1991	91,347	1,300
TOTAL	$431,277	$22,650

By using this five-year recap of sales and earnings, we can address the Information Center budget—that is, should the CEOs and their directors agree to a new industry look that includes the Center. Consider an initial five-year budget in the $80 million range. This is three-tenths of 1 percent of the net after-tax profits of only thirty companies in the industry for this five-year period. And, of course, many more than the companies on this list would support the Center.

In short, the financial commitment to the Center could be looked upon as modest, yet the returns would be almost immediately conspicuous.

Center Governance

In order to function, the director of the Center would have to be responsible to society—its customers—not to industry CEOs. Industry concerns would flow from a CEO Board of Advisors to the Center director. In return, the director would give assignments to the CEOs. These would be based on society's needs and on the views of individual citizen conservationists of issues identified by CEOs as well as others.

For the Center, the Board of Advisors would be limited to CEOs, COOs, and owners of businesses ranging from Groups I through VI. Three committees on this board should suffice: a Center Issues Committee, a Center Staffing Committee, and a Center Budget Committee. As long as the director is fulfilling the Center's responsibility toward society in an exemplary manner, the director would have a free hand.

Center Infrastructure

Sometime before the year 2000, the Center would open, comprised of the following facilities: accommodations for Center staff; library and conference rooms; a large and small auditorium; in-house culinary staff; cafeteria and mess hall; dormitories, rooms, and cottages; indoor and outdoor pool and other fitness and recreational facilities; an arboretum and trails. Ongoing programs utilizing the Center should help defray the cost of operation.

Center Programs

The Center's first priority would be to accommodate the Forest Products Industry, but the facilities would be available and programs would be designed to serve all individuals and groups concerned with forestland conservation and management.

Included would be educators of all age groups—from kindergarten through university; all types of government agencies; environmental organizations; urban forest councils at every level; community service organizations; church groups; and outdoor recreation organizations. In short, the Center would be open to all organized groups as well as to the general public; visitors would be welcome.

One of the principal objectives of the Center would be to assist educators to provide their students and the general public with a better understanding of the resilient and dynamic nature of our renewable resources and to make to the public the compelling case for ecosystem management. This would also help efforts to find common ground.

Programs for the benefit of the Forest Products Industry should include a nine-week boot camp where recruits will be educated in all aspects of the Forest Products Industry universe. Graduates will have learned how the needs of society are fulfilled by working them back through the channels of distribution, from manufacturing plants to the raw material in our renewable forestland ecosystems.

Individuals participating in boot camp would include educators, environmentalists, bureaucrats, politicians, entry-level employees, CEOs, and owners of businesses, but Forest Products Industry recruits would be given first priority.

Other groups with priority access to Information Center programs would be primary producers in such fields as coal, oil and gas, and mining, as well as those involved in ranching and farming. Once the Center is operating, other primary producers could be encouraged to locate in mid-America and open similar information center facilities. It is an idea whose time has come.

Center Dividends

Assume the Information Center is in place and staffed with world-class communicators. What would its contributions be? First, merely announcing that the Forest Products Industry exists would constitute progress. This is worth repeating, as a couple of examples will show. In their 11 January 1993 issue, *Fortune* editors devoted six pages to their Economy/Industrial Forecast detailing "How the Industries Stack Up."[5] Here's the line-up:

Aerospace	Computers
Airlines	Financial Services & Insurance
Autos	Environmental Services
Railways & Trucking	Entertainment
Metals	Banks
Oil & Gas	Clothing
Chemicals	Food Processing
Industrial Equipment	Medical Services
Utilities	Pharmaceuticals
Telecommunications	Travel & Tourism

The Forest Products Industry is conspicuously absent.

Fortune editors in the 19 April 1993 issue on international trade devoted six pages to "How NAFTA Will Help America." What the treaty will do for twelve U.S. industries was detailed with a list of industries as follows:

Agriculture	Industrial Machinery
Automobiles, Trucks	Machine Tools
Automotive Parts	Petroleum, Natural Gas &
Chemicals	Related Services
Computers, Components	Pharmaceuticals
& Electronics	Steel Mill Products
Household Appliances	Textiles, Apparel

Again, the Forest Products Industry, with over 1.4 million employees and a major factor in trade with Canada and Mexico, is not listed.

More of the same was evident in the 15 September 1993 issue

of the *Wall Street Journal* where thirty leading industries were selected to showcase the journal's 1993 All-Star Analysts survey. Obviously, the Forest Products Industry, again missing, was not considered a leading industry by the journal.

But back to the Center: Following the momentous announcement of the existence of the Forest Products Industry, the Center could begin addressing misinformation head-on. For example, when antiforestry advocates peddle fabrications in Europe about North American forestland conservation and initiate consumer boycotts, they would have to compete with the truth in six languages.

Center communicators will cut through perceptions like a hot knife through butter. One of the first perceptions to melt would be that our forestlands are being destroyed. Rather, citizens would soon realize that management of our renewable forestland ecosystems and the American dream are compatible goals. Moreover, once the Center concept starts to function, the common sense of citizen conservationists will have enormous impact.

Center Dividends: Common Sense Over Bureaucratic Bumbling

The success stories of forestland management outlined in this book are not exceptions. They are everyday examples of the positive accomplishments of the last fifty years and the direction we must continue to take in the second century of professional forestry. And the owl story in the private sector of northern California is not an aberration either. Rather, it is a common sense example of managing wildlife habitat.

But in public forestlands, as shown, the owl has become an excuse to curtail harvesting. Scientists in the private sector know how to count owls and care for them in the real world, whereas scientists working in public forestlands seem not to know how to manage either of these tasks. Worse, they are reluctant to look at scientific data that undermine their preservationist dogma.

This of course is not news. Alston Chase in his book, *Playing God in Yellowstone*, gives abundant evidence that critical habitat has been severely damaged by continuous bureaucratic bumbling.[6]

According to Chase, it would be difficult to record the many species of wildlife eradicated from the park since its 1872 dedication. He describes in detail park damage that takes place when the concept of "natural regulation" takes precedence over management. In addition, he details how research was literally banned from the park for most of the years of its existence, and he notes how park officials systematically ignored the modest research that was carried out.

Regarding the urgent need for management and research, Chase reviewed the concern expressed by Secretary of Interior Stewart Udall in the early 1960s. At the time, the Park Service research budget was less than $30,000 per year and there were no full-time scientists. In response to Udall, Chase notes:

> The National Research Council of the National Academy created a committee under the direction of William J. Robbins of the National Science Foundation to evaluate Park Service efforts in science and make recommendations for "a research program designed to provide the data required for effective management, development, protection, and interpretation of the national parks; and to encourage the greater use of the national parks by scientists for basic research." The findings of this committee, known as the Robbins Report, was a devastating indictment of Park Service efforts.

Chase summarizes some of the Robbins report:

> The Robbins committee emphasized that parks cannot be left to run themselves. . . .They urged that wildlife inevitably needs human help, and this task in turn requires our best scientific efforts.

> They stated, "No national park is large enough or adequately isolated to be, in fact, a self-regulatory ecological unit," thus "limitation of herds of elk, supervision of visitors in a park, control of water levels. . . controlled burning. . . are necessary functions

of management if a park is to survive in anything like the condition which meets the purpose for which it was established. . . ."

"This Committee believes that management of national parks is unavoidable," where "management" was defined as "activity directed toward achieving or maintaining a given condition in plant and/or animal populations and/or habitats in accordance with the conservation plan for the area."[7]

This information from *Playing God in Yellowstone* is cited for a couple of reasons. One is an event that triggered Secretary Udall's concern at the time, an event that occurred during the winter of 1961-62 which Chase refers to as "the great elk reduction."

This "elk reduction" was an example of "natural regulation" at its worst. "In six weeks," Chase relates, "the direct-reduction teams had shot more than 4,300 elk. Even for hardened observers, watching these symbols of wild America that only a few decades earlier had been on the verge of extinction treated as so much meat on the hoof was upsetting."[8]

The second reason for citing Chase is the Robbins Report recommendation of the need for controlled burning in Yellowstone, a recommendation that was not followed. Instead, fuel loads were allowed to accumulate under "natural regulation" and the result was the tragic 1988 wildfire.

Chase's book and the Nova program marking the tenth anniversary of the Mount St. Helens eruption expose antiforestry perceptions with detailed evidence. But these exposés are insufficient and have not been disseminated and therefore the misleading rhetoric prevails.

Once the Information Center is functioning, however, truth will have its day. Center programs will disclose the high cost to society of "natural regulation" and other misperceptions.

Information Center Dividends: Global Housing Needs

Center dividends will include market opportunities worldwide. And there are many. Housing is one. As incongruous as it may seem, owls, kangaroo rats, and snails are more deserving of a home than people. The majority of our citizens, even those gainfully employed, can no longer afford a home of their own. And this pales before the problems of the millions worldwide who have never experienced adequate shelter.

One of the great opportunities for the Center to serve societal needs would be its ability to coordinate American technology and expertise to provide shelter. And it is precisely our renewable raw material resource base that will enable us to address this problem. But to do so, we would have to begin to manage our resources in the public sector as we do in the private sector.

Housing goals worldwide could then become a high priority for center staff. But a more innovative approach to providing shelter is required, an approach such as that taken by the Boeing Company, for example. Their success in providing air transportation depends on a system that sees a product through from initial design to reliable performance. Boeing would never have succeeded had they only offered pieces and parts.

Yet, the Forest Products Industry expends millions on pieces and parts that go into a home rather than concentrating on the design and reliable performance of a system of shelter, whether a two-room unit for less than a thousand dollars or a mansion for a million.

There are a number of case histories where the complete system is installed off-shore. The most conspicuous of these are major construction projects where housing is part of the project and thus bypasses local customs and the most difficult hurdle of all, which is financing.

For the Center staff, bringing together workable plans utilizing the available technology and expertise to solve housing problems worldwide would not be difficult. Once again, Center planners would coordinate the work which would be contracted out.

Imagine the benefits of our engineered wood-frame housing units, standing intact amidst collapsed masonry and other inferior building materials in the aftermath of death-dealing earthquakes and hurricanes. Financing, of course, will require innovative thinking and planning. But the concept is good; another example of how the Information Center would serve society.

Center Issues: a Recap

The Information Center is not intended to be just another industry sandbox. Nor should it be looked upon as a substitute for any established industry organization. First and foremost, the Center must respond to societal needs.

Second, the Center concept is a means of providing strategic planning assistance to owners and CEOs of the Forest Products Industry. Center programs will be a boon to senior management to allocate limited resources to each company's vested interest in forest-lands, manufacturing, and marketing.

The Center will, moreover, be a source of true information which can be used by professional resource managers as well as lay persons. The information data base will cover global as well as domestic forestland conservation issues.

If industry leaders will not provide leadership, who will? It would be far better for CEOs to support the concept of the Information Center voluntarily, rather than continue to have their feet held to the fire by antiforestry advocates, the media, members of Congress, and judges.

There is still time to prepare the Forest Products Industry for the challenges of the twenty-first century.

PART III

FORESTLAND CONSERVATION: A GLANCE BACKWARD

6

A CONSERVATION ETHIC FOR TREES AND FORESTLANDS

SOON AFTER EUROPEAN settlers began arriving on these shores, they began to use forest products for shelters, utensils, food, fuel, and much else besides. In the seventeenth and eighteenth centuries, wood was used for ships, homes, churches, and fortifications, as well as schools and commercial buildings, furnishings, tools, weapons, bridges, and transport vehicles. Indeed, without forestland resources during this early period, nearly the entire population would have been, literally, out in the cold.

Samuel T. Dana in his book *Forest and Range Policy* notes how what was to become the Forest Products Industry dealt with the first cargo of wood products shipped from Massachusetts Bay Colony to England in 1623 on the "good ship 'Anne,'" and again in 1626 by a shipment of lumber from New Amsterdam.[1] "By the early 1630s," Dana writes, "several small sawmills were operating in Maine, New Hampshire and New York."

Today, our forestlands provide us with a wide array of products which are essential to our well-being. Only the air we breathe, the water we drink, and the food we eat are more vital to us physically than our trees and forestlands. And, of course, healthy trees and forestlands also contribute significantly to our supply of oxygen, clean water, and the packaging and transportation of food.

There is more, however, much more to our relationship with trees and forestlands than our physical well-being. There is a love of trees and a spiritual affinity that defies explanation. Ironically, this same affinity can become detrimental, making us vulnerable to emotional appeals about forestland stewardship that can do more harm than good.

Even centuries back, the exploitation of our forestlands was of great concern to a few far-sighted citizens. Dana gives one example: "An attempt to prevent widespread clearcutting was made by William Penn, who in 1681, in the document governing the establishment of a colony in 'Penn's Woods,' provided that 1 acre of trees must be left for every 5 acres cleared."[2]

Back in 1875, the fears expressed by some people would seem all too familiar. And for good reason. Wildfires, clearing the land for agriculture, and harvesting trees primarily for lumber were having a devastating effect on forested regions all across the United States. Concerned citizens feared that wasteful practices were destroying their green heritage.

AMERICAN FORESTRY ASSOCIATION, 1875

The ugly consequences of wasteful practices in our forestlands were very much on the minds of a small group that joined with John A. Warder in 1875 to organize the first national citizens conservation association. Much of the colorful history of this movement is detailed in Henry Clepper's *Crusade for Conservation*, the Centennial History of the American Forestry Association.

A few attenuated biographies of these dedicated leaders exhibit the passion for forestland conservation that laid the groundwork for enabling legislation in the first fifty years of the fledgling crusade for conservation. These sketches offer inspiration.

Henry Clepper (1901—1987)

Henry Edward Clepper was born 21 March 1901 in Columbia, Pennsylvania; he graduated from Pennsylvania State Forest Academy in 1921 and began his career with the Pennsylvania Department of Forests and Waters where he remained for the next fifteen years.[3] In 1936 he became an information specialist for the Forest Service in Washington,

D.C. The following year, he was appointed executive secretary of the Society of American Foresters and served in this position and as managing editor of the Journal of Forestry for twenty-eight years, until his retirement in 1966, except for a two-year leave of absence with the War Production Board during World War II.

From 1957 to 1965 he was an advisor to the Forestry Committee of the Food and Agriculture Organization of the United Nations at the FAO council's biennial conferences in Rome.

Clepper authored more than one hundred articles and bulletins on forestry and natural resources. His writings contributed more to the history of forestland conservation than those of any other individual. His books include: "American Forestry-Six Decades of Growth," 1960; "The World of the Forest," 1965; "Origins of American Conservation," 1966; "America's Natural Resources," 1967; "Forestry Education in Pennsylvania," 1967; "Leaders in American Conservation," 1971; "Professional Forestry in the United States," 1971; "Crusade for Conservation," 1975; "Famous and Historic Trees," 1976 and "Careers in Conservation," 1963 and 1979.

Clepper was elected a fellow of the Society of American Foresters and in 1957 was the recipient of the Gifford Pinchot Medal. From the American Forestry Association, he received the Distinguished Service Award in 1970 and the John Aston Warder Medal in 1977. Clepper died on 26 March 1987 in Washington, D.C.

In 1875, Clepper wrote, "Dr. Warder issued his call for a conference on forestry in Chicago to fellow horticulturists, nurserymen, botanists, and ordinary interested citizens."[4] This landmark conference took place in the Grand Pacific Hotel on 10 September, which was the opening date of the Chicago Exposition of 1875 and followed a meeting of the American Pomological Society, of which Dr. Warder was a vice president.

The conference was attended by twenty-five prominent citizens from various states and territories who took it upon themselves to breathe life into a conservation ethic for trees and forestlands. At this organizing meeting, Dr. Warder was elected the first president of the American Forestry Association, a position he held for seven years.

Dr. John A. Warder (1812—1883)

John Aston Warder was born on 19 January 1812 in Philadelphia, graduated from Jefferson Medical College in 1836, and practiced medicine in Cincinnati until 1855.[5] While still an active physician, he edited "Western Horticulture Review" and "Botanical Magazine." He was a member of the Ohio State Board of Agriculture and the Ohio Fish Commission.

Dr. Warder was interested in medical science, astronomy, meteorology and the study of plant life. He was a widely recognized expert in fruit culture. In 1855 he purchased a farm near North Bend, Ohio, in order to pursue his horticultural avocation on a full-time basis.

In 1873 Dr. Warder was appointed United States commissioner to the International Exhibition in Vienna and authored the commission's official report on forests and forestry. In it he listed the European forest schools dating back to 1813 and noted the European practice of holding international congresses of forestland managers. This was at a time in the United States when there was not a single forester nor any professional organization concerned with forestland conservation. Dr. Warder died on 14 July 1883.

The second Forestry Association meeting was held during the Centennial Exhibition in Philadelphia on 15 September 1876. The invitation for this meeting stated that the association's purpose was "the fostering of all interest of forest-planting and conservation on this continent."[6]

At the meeting a constitution was adopted that included the following objective: "the protection of the existing forests of the country from unnecessary waste, and the promotion of the propagation and planting of useful trees."[7]

Interestingly, Gifford Pinchot (about whom more in due course) apparently convinced himself that he had launched the conservation movement. In his autobiography, *Breaking New Ground*, he recounts how the "concept of conservation flashed through my head" while riding his horse in Washington, D.C.'s Rock Creek Park on a winter day in 1907.

Whatever Pinchot may have believed, in truth he was a latecomer. William N. Sparhawk in an article, "History of Forestry in America," points out that the term "Forest Conservation was in use more than 30 years before it was taken up and popularized by Gifford Pinchot."[8]

Dr. Warder's chief concern was the conservation of forestland resources while Pinchot's main wish was to conserve all natural resources. There is no question, however, that Dr. Warder and the American Forestry Association were instrumental in launching the conservation movement in the United States.

Gifford Pinchot (1865—1946)

Gifford Pinchot was born on 11 August 1865, in Simsbury, Connecticut. He graduated from Yale University in 1889, after which he studied forestry for thirteen months at the National School of Waters and Forests in Nancy, France.[9]

Pinchot was the first American citizen to receive forestry instruction and is credited with establishing the first systematic forestland management program in 1892 at the Biltmore Forest in North Carolina. In 1898, he was appointed chief forester in the Division of Forestry in the Agriculture Department where he served until dismissed by President Taft in 1910 as a result of the Ballinger-Pinchot controversy.

In 1900 Pinchot founded the School of Forestry at Yale University. That same year he also founded the Society of American Foresters with seven charter members. He became its first president and was elected a fellow in 1918 and received the Sir William Schlich Memorial Medal for distinguished service to forestry in 1940.

Pinchot was active in the American Forestry Association, serving as chairman of its executive committee in 1899; nearly half a century later, in 1943, after taking issue with the association's policy favoring cooperative forestland conservation as opposed to his own preference for federal regulation, he tendered his resignation.

In 1908, Pinchot founded the Conservation League of America, which was renamed the National Conservation Association a year later. Within this organization, he pursued his personal agenda for conservation issues until 1923, when it was merged with the American Forestry Association.

Pinchot ran unsuccessfully for the United States Senate from Pennsylvania in 1920 and 1926. He was commissioner of the Pennsylvania

Department of Forestry, 1920-22, and governor of Pennsylvania for two terms—1923-27 and 1931-35. He authored four books and numerous papers and reports on conservation subjects and received honorary degrees from five universities. Pinchot died 4 October 1946.

Another important advocate of forestland conservation who was well ahead of Pinchot, Dr. Franklin B. Hough, was the first Forestry Agent named by the federal government. Clepper notes in *Crusade* that Dr. Hough's awareness of the depletion of the nation's forestland resources came about as a result of his work with the census.

Dr. Franklin B. Hough (1822—1885)

Franklin Benjamin Hough was born on 22 July 1822, in Martinsburg, New York, graduated from Union College in 1843, and obtained his M.D. at the Western Reserve Medical College in 1848.[10] He was physician and surgeon for a New York regiment during the Civil War. In 1854 and 1865 he was director of the New York State census and superintendent of the United States census in 1870.

Clepper relates that Dr. Hough's 1873 address, "On the Duty of Governments in the Preservation of Forests," before the American Association for the Advancement of Science in Portland, Maine, was "the most influential pronouncement in behalf of forest conservation made in America up to that time."[11]

Dr. Hough was named the first federal Forestry Agent in the Agriculture Department in 1876 and became the first Chief of Forestry in 1881, serving until he was succeeded by Nathaniel Egleston in 1883.

Dr. Hough was the first treasurer of the American Forestry Association and later served as its recording secretary. He was a fluent writer of papers and speeches and authored four official Forestry Reports in 1877, 1880, 1882, and 1885. He also wrote <u>Elements of Forestry</u> in 1882, the first book on practical forestry to be published in the United States. Hough's huge 618-page <u>Report upon Forestry</u> published in 1877 was so significant that Congress authorized the printing of 25,000 copies.

Dr. Hough was awarded a special diploma of honor by the International Geophysical Congress in Vienna in 1882, in recognition of his tireless work promoting professional forestry. After being replaced as chief forester, Dr. Hough continued to champion forestland conservation and management.

In 1885 Dr. Hough, along with Bernhard Fernow, drafted the legis-lation that lead to the creation of New York State's Forestry Commission and he worked as an agent in the Division of Forestry until his death on 11 June 1885.

Clepper also mentions the American Forestry Council, which was another citizens' group "desirous of concerted action for forest protection."[12] Its beginning stems from a gathering in 1873 of the Farmer's Club, a unit of the American Institute, based in New York City, where a committee was formed to promote the council in Congress, but the independent effort was short-lived. The committee took the name American Forestry Council and called for a National Forestry Convention to be held in Cape May, New Jersey, in September 1876. Only a few individuals responded, and two weeks later, the American Forestry Council decided to merge with the American Forestry Association.

AMERICAN FORESTRY CONGRESS, 1882

The American Forestry Association merged with the American Forestry Congress in August 1882. Dr. Warder stepped down as president of the American Forestry Association at the Rochester, New York, meeting 29 June 1882. He was succeeded by Dr. George B. Loring of Salem, Massachusetts, a former congressman and at that time U.S. commissioner of Agriculture. Dr. Loring was also elected president of the American Forestry Congress at its first meeting in Montreal that August, when the association was joined with the American Forestry Congress.

The first American Forest Congress was organized in Cincinnati in April 1882, independent of the conservation efforts of the American Forestry Association. This landmark event was actually triggered by a group of visitors from outside the country. Concerning this ambitious conservation effort, Clepper has unearthed a "curious genesis" dating back to a Revolutionary War hero:

In the autumn of 1881, the Government of the United States celebrated the centennial anniversary of the surrender of Lord Cornwallis in Yorktown, Virginia. Seven Germans bearing the name von Steuben, representing the family of Baron Friedrich von Steuben, the Prussian general who served the Continental Army so loyally and effectively, were invited to participate.

Six were officers in the Prussian Army; the seventh, Richard Baron von Steuben, a former Army officer, was Oberforster (head forester) in the Prussian Forest Service. After the ceremonies at Yorktown, the Germans visited other historic places in the United States, among them Cincinnati, where citizens held a reception for them.

Oberforster von Steuben, having observed the waste and neglect of American forests, commented that forest conservation was a subject urgently needing public attention.

Among the citizens challenged by this comment was William L. DeBeck, a newspaperman of the city. In January 1882, Colonel DeBeck and other concerned individuals, seeking ways to arouse public interest, announced an open meeting in the Gibson House.

Committees were formed and it was decided to hold a national convention of five days duration in the city, beginning April 25, 1882.

The governor proclaimed Arbor Day on April 27 to be a public holiday, and tree planting ceremonies were carried out in Eden Park. Memorial groves were planted in honor of authors, soldiers, statesmen, and other distinguished persons. On hand to witness the exercises were 50,000 spectators.

Open meetings were held in Springer Music Hall. More than one hundred addresses and papers were presented during the sessions. John A. Warder's enthusiasm for trees and tree planting is shown by his participation on the program with six separate papers, ranging in subject matter from individual tree descriptions to species mixtures in plantations.[13]

The new name of the American Forestry Association, the American Forestry Congress, was retained until 1890, when it was once again changed to the American Forestry Association. This name was incorporated in 1897.

According to Clepper, "A list of members published in the 'Proceedings' contained 266 names; 92 of these were members of the American Forestry Association before consolidation."[14] Others besides Dr. Warder and Dr. Loring who served on the organizing committee of the new American Forestry Congress were Dr. Hough and General C. C. Andrews, a forestland conservation pioneer in Minnesota.

Bernhard Fernow, a principal architect of the forestry profession in North America, played a leading role in bringing about the consolidation.

Bernhard Fernow (1851—1923)

Bernhard Eduard Fernow was born on 7 January 1851, in the Province of Posen, Prussia; he studied law at the University of Königsberg and completed a forestry curriculum at Hanover-Muenden Academy in 1873. He was licensed and served in the Prussian Forest Service until he emigrated in 1876 to the United States.[15]

From 1878 to 1885, Fernow was employed by Cooper Hewitt & Company to bring a 15,000-acre Pennsylvania forest under professional forestry management. He soon became an active force in the public forestry movement.

In 1886 Fernow was appointed chief of the Division of Forestry in the Agriculture Department under President Grover Cleveland, where he served until 1898. He quickly assembled a skilled staff and commenced research in silviculture, pathology, wood technology, forest products, and tree planting on the Great Plains.

Fernow emphasized that forestry meant management and when

properly carried out should be economically practical. His administration of the Forestry Division and his many circulars, bulletins, articles, and addresses laid the groundwork for forestry in North America.

His legal training in Germany enabled Fernow throughout his career to draft model bills for both federal and state legislatures. Much credit is due him for the Forest Reserve Act of 1891 and the Forest Management Act of 1897.

Fernow championed the cause of forestry education and research throughout North America. From 1898 to 1903 he was director of the New York State College of Forestry at Cornell University where he organized the first professional forestry curriculum in the Western Hemisphere.

In 1907 he helped establish the Department of Forestry at Pennsylvania State College. That same year he became the first dean and organized the faculty of forestry at the University of Toronto, from which he retired in 1919.

While at Cornell in 1902, Fernow established the Forestry Quarterly which he edited from 1903 to 1916. When the quarterly became the Journal of Forestry after merging with the "Proceedings" of the Society of American Foresters, he became editor-in-chief, working in that capacity from 1917 to 1923.

Fernow served as secretary of the American Forestry Association from 1883 to 1895 and as chairman of its executive committee. In 1895 he was elected a vice president of the American Association for the Advancement of Science. He served as president of the Society of American Foresters in 1914 and 1916, and was elected a Fellow in 1918.

Fernow helped organize and was first president of the Canadian Society of Forest Engineers in 1908. From 1910 to 1923 he served on the Canadian Conservation Commission. He received honorary LL.D. degrees from three universities and authored three books, which became standard texts, and over 250 papers and bulletins.

Throughout his long career he was a preeminent professional forester, administrator, educator, author, and editor. Fernow died on 6 February 1923.

At the August 1882 inaugural meeting of the American Forestry Congress in Montreal, Fernow led the preparation of a new constitution. Its objectives were: The discussion of subjects relating to tree planting; the conservation, management, and renewal of forests; the climatic and other influences that affect their welfare; the collection of forest statistics, and the advancement of educational, legislative or other measures tending to the promotion of these objects.[16]

The new constitution was a critical step forward in the conservation ethic for trees and forestlands. In *Crusade*, Clepper observes that "*management* was equated with *conservation* and renewal" [emphasis his]. And that "Fernow's technical education and field experience doubtless prompted him to emphasize the importance of management, because he knew, even if others present did not sense the significance of the term, that woodland preservation alone was not forestry."[17] Fernow appeared to be aware that the conservation ethic they were nurturing went well beyond leaving forestlands in their "natural" state.

AMERICAN POMOLOGICAL SOCIETY, 1852

Clepper also credits a number of horticulturists and nurserymen as originators of the forestland conservation movement and the development of a conservation ethic for trees and forestlands. Most of the twenty-five founders of the American Forestry Association were members of the American Pomological Society, which was organized at a conference in Buffalo, New York, on 1 September 1848.

This conference was hosted by the New York Agriculture Society and was called the North American Pomological Convention, soon and briefly to become the American Pomological Congress. In 1852, when a constitution was adopted, "the name was changed to the present American Pomological Society. . . the first national association in America to foster the cultivation and development of the good fruits of the land."[18] It is active to this day with over one thousand members.

AMERICAN ASSOCIATION OF NURSERYMEN, 1876

Following the 1875 inaugural meeting of the American Forestry Association in Chicago, sixty-two nurserymen also met in Chicago on 14 June 1876 where they organized the American Society of Nurserymen, Florists and Seedsmen. Elisha Moody of Lockport, New York, was the first president, to be succeeded in 1878 by Ohio's J. J. Harrison.

Both men were also among the founders of the American Forestry Association. Other charter members of the two associations were Robert Douglas of Illinois, Thomas Meehan of Pennsylvania, John Saul of Washington, D.C., and Dr. Warder. This group became the American Association of Nurserymen in 1887, which is active to this day, headquartered in Washington, D.C., with a membership of some four thousand.

Clepper notes in *Crusade*, "In sum, they were good and useful men whose signal achievement it was to start three enduring associations which have contributed much to the social welfare and the improvement of the environment."[19]

There are several striking aspects in the early history of the conservation movement. First, it was indeed a citizens' movement. Though they were few in number, their vision and dedication to a conservation ethic for trees and forestlands was impressive; this helped focus concern for forestland conservation, management, renewal, and wise use.

Second, for the times, a fair degree of science was involved, as well as an interest in advancing a healthy business climate and furthering technical knowledge.

And third, the number of physicians who played key roles is impressive, doubtless in large measure because of their awareness of the relationship between healthy forests and healthy people.

GEORGE PERKINS MARSH: WORDS OF WISDOM

These citizen conservationists were probably familiar with and influenced by the observations of George Perkins Marsh (1801-1882) in his book, *Man and Nature*, published in 1863. Davis describes this book as "the first treatise on environmental history," because it dwells at length on the ecology of forestlands and consequences of deforestation.[20]

Marsh was a scholar, statesman, diplomat, and author, who had spent a good deal of time in Europe and the Middle East. In Europe he saw firsthand how deforestation and grazing had ravaged the land.

To avoid this, he urged the American people to act on their own and not wait for governments to take the initiative.

Marsh's book helped give citizens a new awareness of their environment and shaped the thinking of conservationists throughout the nineteenth and twentieth centuries. He devoted great attention to trees and forestlands and obviously favored healthy forestland as much as he abhorred deforestation.

The revised edition, entitled *The Earth as Modified by Human Action,* was published in 1874. While he refers to foresters, there is no mention of the profession of forestry; rather, he covers the subject under the heading "The Economy of the Forest."

In this section, Marsh unmistakably supports the management of forestlands rather than allowing nature to take its course. For example, he argues:

> Notwithstanding the difference of conditions between the aboriginal and the trained forest, the judicious observer who aims at the preservation of the former will reap much instruction from the treatises I have cited, and I believe he will be convinced that the sooner a natural wood is brought in the state of an artificially regulated one, the better it is for all the multiplied interests which depend on the wise administration of this branch of public economy.[21]

Unfortunately, citizen conservationists who wished to share their concern for trees and forestlands in a positive way were few in number. And fewer, no doubt, were aware of Marsh's words of wisdom. Consequently, the general public did not become involved in the 1875 Crusade for Conservation, and the exploitation of our nation's forestlands continued for the better part of the next seventy-five years.

The early conservationists could hardly have anticipated the myriad complexities and general confusion that were later to arise on the subject of a conservation ethic for trees and forestlands, nor the magnitude of the vast effort to educate the public about the many intricate matters dealing with forestland conservation and our environment.

Fortunately, education was one of their first priorities. The early leaders knew then as we know now that concerned citizens must be made aware of the value of the forestland resource base and that healthy trees are essential for our health and well-being. An educational program was of paramount importance to instill in the mind that each and every individual can make a difference.

7

TREES AND EDUCATION

WELL BEFORE DR. John A. Warder founded the American Forestry Association in 1875, he had done a great deal in the way of education by imparting to the public his concern for trees and forestlands. And education, of course, was a cornerstone of the association. Samuel T. Dana says it best in his forward to Henry Clepper's *Crusade for Conservation*:

> The Association's chief tool for attaining its objectives has been education—of the general public, schoolchildren, forest owners, bureaucrats, and legislators, notably at the federal level.
>
> Much reliance has been placed on its magazine and other publications. Constructive use has been made of its annual meetings, of special meetings, such as its five Forest Congresses and its 1972 conference on tree planting; and on field campaigns such as the Southern Forest Fire Project.
>
> Although not a lobbying organization, it has kept the Congress informed of its views on subjects within its field of competence. In short, it has regarded education as in the long run the most effective way of spreading the gospel.[1]

Thus, for some 120 years, education has been a primary goal of the American Forestry Association and it is likely to remain so.

The difficulty faced by the early leaders of forestland conservation was the lack of a professional forestry curriculum in the United States. By contrast, in Europe, forestry had long been practiced; by the late eighteenth century, the profession was well established.

As noted earlier, Dr. Warder in the course of his 1873 trip to Vienna was one of the first Americans to observe the benefits of professional forestry. Shortly thereafter, the "Father of Forestry" in Pennsylvania, Dr. Joseph Rothrock, was instrumental in introducing Old World forestry education to these shores.

In 1877, Dr. Rothrock was appointed F. Andre Michaux lecturer in forestry methods at the University of Pennsylvania. After studying botany in Strasbourg and observing the well-managed German forestlands, he wrote a prize-winning essay, "Forestry in Europe and America."

Dr. Joseph Rothrock (1839—1922)

Joseph Trimble Rothrock was born on 9 April 1839, in McVeytown, Pennsylvania, attended Freeland Seminary (now Ursinus College), and studied under Asa Gray at Harvard's Lawrence Scientific School.

His education was interrupted by the Civil War. After Rothrock, a captain in the 20th Pennsylvania Cavalry, was wounded at Fredericksburg, he returned to Harvard, completed his undergraduate degree in 1864, and subsequently earned an M.D. from the University of Pennsylvania in 1867.

Prior to starting a career in medicine in 1869, he taught botany at Pennsylvania State Agricultural College. In 1873, Dr. Rothrock left his medical practice to serve two years as surgeon and botanist for the United States Geological Survey.

On his return, he was appointed professor of botany on the faculty of medicine at the University of Pennsylvania where he served, aside from his European sojourn, until retirement.

After his European experience, Dr. Rothrock was offered a position at Harvard, but he declined in order to concentrate on forestland conservation through lectures and public education in Pennsylvania.

On 30 November 1886, Dr. Rothrock helped organize the Pennsylvania Forestry Association and was elected its first president. When the Pennsylvania Department of Agriculture, Division of Forestry, was established in 1895, he was appointed the first commissioner of forestry.

Dr. Rothrock was instrumental in acquiring state forestland reservations and establishing the State Forest Academy at Mt. Alto in 1903. When he retired in 1904, the state forestlands comprised more than 440,000 acres.

He also served as vice president of the American Forestry Association, participated in the American Forest Congress in 1905, and was a delegate to the Conference of Governors and the Joint Conservation Congress in 1908.

His achievements as the "Father of Forestry" in Pennsylvania were recognized in 1915 when the Society of American Foresters made him an honorary member. Dr. Rothrock died on 2 June 1922.

Another United States citizen who closely observed professional forestry in Europe was Austin F. Cary. His college education was in botany and entomology, but his interest focused on practical forestry, a study he followed in 1890 after graduation.

In pursuit of this interest, Cary traveled to Europe several times, concentrating on the Black Forest of Germany. William B. Greeley, who was to become chief forester in 1920, called Austin Cary "forester-extraordinary of the north woods." Greeley met Cary in the snowbound north woods of New Hampshire in the winter of 1902, at the beginning of his own career. They would become lifelong friends.

Austin Cary (1865—1936)

Austin F. Cary was born 31 July 1865, in East Machias, Maine; he majored in science at Bowdoin College with an emphasis on botany and entomology, receiving his A.B. degree in 1887 and M.A. three years later. He also studied at Johns Hopkins and Princeton. Prior to his employment by Berlin Mills Company in New Hampshire in 1898—the first company forester in America—he did extensive freelance work in New England forestlands.

Cary's early writings, dealing with improved forestland use, thinning, tree planting, and selective cutting, caught the attention of many in the Forest Products Industry, few of whom agreed with Cary. But opposition merely increased his missionary zeal, and finally launched Cary on a career in writing, speaking, and traveling for his cause.

Cary's early research on selective cutting methods, tree growth, entomology, and other silviculture practices was published by the Maine Forestry Service and _Paper Trade Journal_. While on the faculty at Yale (1904-05) and at Harvard (1905-09), he wrote his only book, _A Manual for Northern Woodsmen_, published in 1909.

After a brief stint as superintendent of forestry in New York State, Cary joined the Forest Service in 1910 and from 1917, until he retired in 1935, his field work was concentrated in the Southern pine region.

In the Forest Service, Cary's official title was logging engineer, but he actually functioned as a roving extension forester. Though his style was blunt and colorful, it was based upon broad experience. His impact, particularly in the South, was immense. He was so successful that by the 1920s he had become known as the "Father of Southern Forestry."

Like Fernow and Greeley, Cary correctly perceived that once private forestland owners realized the possibility of long-term continuous profits, they would shift from exploiting to renewing their forestlands.

He consistently demonstrated the value of the improved forestry practices he had observed during his travels in Germany. His tireless work with forestland owners large and small laid the groundwork for industrial forestry as we know it today. On his retirement, Cary returned to Maine where he died on 28 April 1936.

Though Dr. Franklin B. Hough, the first forestry agent named by the federal government, did not advocate technical training in forestry, he became a key player in bridging the forestry education gap between Europe and the United States.

In 1877, as special agent appointed by the commissioner of agriculture, Frederick Watts, Dr. Hough's task was to investigate the consumption of forest products and to recommend the best means of forestland conservation and renewal. In the course of this assignment, he traveled extensively in the United States, Canada, and Europe.

Dr. Hough reviewed as much written material as he could find and corresponded with anyone who could possibly be of assistance. Fernow considered Hough's 1877 Report upon Forestry "by far the best and most useful publication in forestry in this country."[2] Gifford Pinchot referred to Dr. Hough as "perhaps the chief pioneer in forestry in the United States." Sir Dietrich Brandis, an internationally known German authority, also had high praise for Dr. Hough and his work.

In 1881, when Dr. Hough became chief of the Division of Forestry in the Agriculture Department, he spent the summer of that year in Europe gathering information about forestry practices and speaking with prominent forestry leaders including Brandis.

As noted earlier, Gifford Pinchot was the first American citizen to receive forestry instruction in Europe. He attended the National School of Waters and Forests in Nancy, France, from December 1889 to December 1890. He also toured forestlands in Switzerland and Germany.

Henry Graves followed Gifford Pinchot's example, studying forestry for two years at the University of Munich under Heinrich Mayr. During this time, he, like Dr. Hough, became well acquainted with Sir Dietrich Brandis.

Henry Graves (1871—1951)

Henry Solon Graves was born on 3 May 1871 in Marietta, Ohio. He graduated from Yale in 1892, after which he studied at the University of Munich. In 1898, he was appointed assistant chief in the Division of Forestry, Department of Agriculture, under Pinchot. He resigned in 1900 to become dean of the newly established Yale School of Forestry.

In 1910, Graves returned to the Forest Service as its chief and served until 1920. During his tenure, National Forest land decreased from 172 million acres to 154 million, but this modest decrease was a tribute to Dean Graves' competency as an administrator. Many advocated drastically reducing Forest Service holdings after Pinchot was dismissed.

During World War I, Dean Graves took military leave to assist in organizing production of forest products for the Corps of Engineers, attaining the rank of lieutenant colonel. He resumed his deanship at Yale in 1922 and remained in that capacity until his retirement in 1939.

Dean Graves was recognized as the nation's foremost forestry educator. He became a director of the American Forestry Association in 1900 and was elected president in 1923 and 1924, but resigned because of his demanding responsibilities at Yale. In 1934, when Association President George D. Pratt was forced to leave for health reasons, Dean Graves succeeded him and was reelected in 1935 and 1936.

Graves was a charter member of the Society of American Foresters, served as president in 1912, was elected a fellow in 1918, was awarded the Sir William Schlich Memorial Medal in 1944, and was first recipient of the Gifford Pinchot Medal in 1950. He was also awarded honorary LL.D. degrees from three universities. He authored numerous articles, bulletins, and monographs as well as two pioneer textbooks. Dean Graves died on 7 March 1951 in New Haven, Connecticut.

A few European professional foresters visited the United States, and some became residents, notably Bernhard Fernow and Dr. Schenck. Dr. Schenck arrived in the spring of 1895, to assume the position of resident forester at George W. Vanderbilt's Biltmore estate near Asheville, North Carolina, and made one of the most significant contributions to professional forestry.

Dr. Carl A. Schenck (1868—1955)

Carl Alwin Schenck was born in Darmstadt, Germany, on 26 March 1868. He studied forestry at the Institute of Technology in Darmstadt, the University of Tubingen, and the University of Giessen, where he earned his doctorate in 1895.

While a student at Giessen in 1889, he met Sir Dietrich Brandis and Sir William Schlich; for several years he assisted them on study tours of forestland practices in France, Germany, and Switzerland, conducted on behalf of British Empire forestry students.

In 1895, as noted, he left Germany for his Biltmore forest responsibility where he remained until dismissed by Vanderbilt in 1909. On 1 September 1898, Dr. Schenck established the Biltmore Forest School, a month in advance of the New York State College of Forestry at Cornell.

The Biltmore site, now within the Pisgah National Forest, was commemorated in 1968 as the Cradle of Forestry in America. Dr.

*Schenck continued to operate the Forest School until 1913, annually con-
ducting study tours to forestlands throughout the United States and to
Europe every other year.*

*Following his return to Darmstadt in 1913, he served as an officer
in the German army during World War I. After the war, he worked as a
forestry consultant and continued to lead tours for visiting forestry students
in addition to his lecture and teaching assignments in the United States. In
the aftermath of World War II, Dr. Schenck was employed by the American army of occupation as chief forester for Greater Hesse.*

*On 7 July 1951, the Weyerhaeuser Company dedicated the Carl
Alwin Schenck working circle on its Millicoma Tree Farm near Coos Bay,
Oregon. In 1952, Dr. Schenck received the honorary degree, doctor of
Forest Science, from North Carolina State University. He died in Germany on 15 May 1955. In 1988, the Society of American Foresters
established the Carl Alwin Schenck award for outstanding forestry educators who exhibit the qualities of teaching exemplified by Dr. Schenck.*

BILTMORE FOREST SCHOOL, 1898

Dr. Carl Schenck initiated formal forestry education in the
United States, it would appear, by happenstance. Shortly after his
arrival in 1895, Overton W. Price, a young college student from the
University of Virginia, who was employed on a nearby estate, became
interested in becoming a forester. He probably had heard of Pinchot's
work at Biltmore prior to Dr. Schenck's arrival. Since there were no
American schools of forestry, he asked Dr. Schenck for permission to
work for him as an unpaid apprentice. Dr. Schenck agreed and the
following year accepted a second student, Edwin M. Griffith.

Price then went on to study forestry under Brandis for two years
at the University of Munich. Before returning to the United States, he
worked briefly as a forester in Europe, during which time he continued
his close association with Brandis. In 1899, Price joined the Forest Service and became Pinchot's trusted deputy and second in command. In
later years, Griffith would become state forester of Wisconsin.

By 1898, Dr. Schenck recognized the need for a more formal-
ized curriculum and published his first catalog. From then on, an

increasing number of applicants was offered a one-year course in forestry. Classroom courses in theory were supplemented with practical field instruction—every second year Dr. Schenck would lead his students on a tour of European forestlands—and when satisfactorily completed, a degree of Bachelor of Forestry was awarded.

Clepper's *Professional Forestry in the United States* sums up the Biltmore Forest School story[3]: "In all, from 1898 until 1913 there were some 350 students. Dr. Schenck shared instruction with visiting educators and scientists whose lectures included forest entomology and pathology, geology, and wood utilization. Clepper also noted Dr. Schenck's emphasis on training foresters to work in industry. In this he was far ahead of his time; the demand for industrial foresters would not be significant for several decades."

SIR DIETRICH BRANDIS: A MAN OF VISION

In his book, *The Birth of Forestry in America*, Dr. Schenck reveals a curious historical footnote to the story of American forestland conservation. At the age of eighty-three, he cited the influence of his early mentor, Sir Dietrich Brandis, as being profoundly important in bridging the educational gap between Europe and America.

In Dr. Schenck's words, "Sir Dietrich might be called the 'grandfather of American forestry,' although he never visited the United States. Undeniably, Gifford Pinchot, Overton Price, Henry Solon Graves, Frederick E. Olmsted, Austin Cary, and Edwin M. Griffith, all of whom were important in the early forestry movement, were decisively influenced by him, as was I in particular."[4]

(Olmsted, a cousin of Frederick Law Olmsted, Jr., and another important figure in the annals of forestry, graduated from Yale in 1894, studied at Biltmore, and subsequently with Brandis at the University of Munich. He then took on the responsibility for several important assignments in the Forest Service under Pinchot including resolution of boundary location procedures for public forestland, as well as field inspection operations.)

Sir Dietrich Brandis had a profound impact on forestland conservation and management in America, largely as a teacher. He seldom missed an opportunity to compare notes with anyone who knew something about the subject. His keen interest in education and his dedication to professional forestry provided a rare vision of what might be accomplished in a land he never visited. In one of some thirty letters he wrote to Dr. Schenck between 1895 and his death in 1907, for example, he observed:

> There are two powerful interests at work in the U.S. in favor of conservative forest management:
>
> I. The wish of lumbermen to adopt conservative lumbering.
>
> II. The millions of capital invested by paper pulp manufacturers in machinery.
>
> If I do not mistake, these two interests will do more for the cause of forestry than either the federal government or the governments of the individual states. I will except Pennsylvania where, under Professor Rothrock, matters seem to be progressing on correct lines. [Bonn, 2 December 1898]

The insights expressed were certainly accurate as the passing years would prove. Brandis' contributions to American professional forestry were recognized by the Society of American Foresters in 1903, when he was elected the first honorary member of the Society.

PROFESSIONAL FORESTRY INSTRUCTION BEGINS

While the United States lacked formal schools of forestry in the late nineteenth century, enough interest was being generated to attract both teachers and students to a wide range of technical courses related to forestry.

Forestry lectures were first offered at Cornell in 1873 and at Yale the following year. During the 1880s, college-level courses were made available on a good many campuses—in Michigan, New Hampshire, Massachusetts, Iowa, Missouri, North Carolina, and Pennsylvania. In the Far West, the first general forestry course was listed in the 1894 University of Washington catalog. The offerings were usually taught by botany or horticulture faculty members who emphasized tree planting and the agricultural aspects of the subject.

Cornell University's four-year forestry curriculum, the first of its kind, had been part of New York Governor Black's 1898 platform. The nation's chief forester, Bernhard Fernow, was invited to Cornell to develop the course of study.

The resulting program, however, was short-lived. In 1903 Governor B. B. Odell, Jr., vetoed the appropriation and the Cornell School of Forestry was forced to close its doors. The reason for the veto sounds all too familiar—Governor Odell acted "in response to protests from owners of recreational camps who opposed the logging conducted on the school forest."[5] Professional forestry bowed to political pressure.

But soon things looked up. Encouraged by "an endowment from the family of Gifford Pinchot" Yale University commenced its graduate program in forestry in 1900. The Yale School of Forestry is thus the oldest continuous program in the country.

Yale's two-year course of study was followed in 1903 by a four-year curricula at Michigan State College, the Pennsylvania State Forest Academy at Mont Alto, and the universities of Michigan, Maine, and Minnesota. Mont Alto was established by the Pennsylvania legislature with a specific charge—to train foresters to manage the Commonwealth's forestland reserves.

In 1904, forestry schools began operation at Harvard, Iowa State College, and the University of Nebraska, although the Nebraska school didn't last. The same was true with Colorado College, whose program, begun in 1905, was discontinued in 1934. In 1906, Oregon State College and the University of Georgia announced offerings in forestry programs, to be followed the next year by Pennsylvania State College, the University of Washington, and Washington State College. In 1909, the University of Idaho enlisted in the cause.

In December 1909, the first national forestry education conference was convened in Washington, D.C., with Henry Graves serving as chairman. A committee of five was appointed to study the kind of technical education and standards of instruction needed to assure competent forest school graduates.

A second National Conference in December 1920 led in 1933 to the Society of American Foresters standards for accredited curricula, which were implemented two years later. Twelve institutions were granted accreditation in the first year: Yale, Michigan, Minnesota, Michigan State, Iowa State, Oregon State, Pennsylvania State, Washington, and Idaho mentioned earlier, along with the universities of California and Montana and the state university of New York at Syracuse.

Today there are forty-six institutions with accredited curricula; in addition, twenty-two schools offer approved technical forestry programs (all listed in the appendix). The society's accreditation process provides for reviews of institutional objectives, curriculum, faculty, students, administration, parent institution support, and physical resources and facilities. Each school receives a written review every five years and a formal on-site visitation every ten years.

FORESTRY AS A PROFESSION

Much credit is due Gifford Pinchot for establishing forestry as a profession in the United States. In addition to endowing the Yale School of Forestry, he was instrumental in organizing the first national forestry education conference while still chief forester.

The Society of American Foresters was founded on 30 November 1900. Pinchot, a leading force in its founding, was elected its first president, serving until 1909, when he was succeeded by his lieutenant, Overton Price. Pinchot was reelected president in 1910 and 1911.

At the first meeting of the society, charter members besides Pinchot were Edward T. Allen, Henry S. Graves, William L. Hall, Ralph S. Hosmer, Overton W. Price, and Thomas H. Sherrard.

At a subsequent meeting in December, eight more foresters were elected to membership: Horace B. Ayres, Bernhard E. Fernow, Edwin M. Griffith, Frederick E. Olmstead, Filibert Roth, Carl A. Schenck, George B. Sudworth, and James W. Toumey. Together they comprised an early Who's Who of American forestry.

The profession of forestry faced enormous problems at the turn of the century. But they pale alongside what the profession currently faces.

Today, instead of a common vision for the future of forestland conservation and management, we have endless confrontation. And endless confusing signals from scientists, educators, environmentalists, and professional foresters. Is it possible to recapture a common vision and purpose in such a contentious climate? The profession and educators are hard at work trying to do just that.

DENVER FORESTRY EDUCATION SUMMIT: NOVEMBER 1991

The Society of American Foresters and the National Association of Professional Forestry Schools and Colleges (NAPFSC) are aware of the many conflicts surrounding forestland conservation and management. Because of their concerns, the two organizations sponsored a national forum on forestry education in Denver in November 1991 and called it "The National Education Symposium on Forest Resource Management in the 21st Century: Will Forestry Education Meet the Challenge?"

The scope, planning, and execution of this event brought to mind the Joint Conservation Conference of 1908, organized by Gifford Pinchot and President Theodore Roosevelt. There were some two hundred participants representing all of the forested regions of the United States, about half of whom were from the various schools of forestry. The other half included representatives from the Forest Service, other agencies, associations, the Society of American Foresters, and the Forest Products Industry.

After the welcoming remarks, four speakers focused on the complex issues confronting forestry education, challenging the audience to confront the task at hand. Participants were then divided into ten workshop subgroups for the remainder of the conference.

At the end of each day, the recorders for the ten subgroups compared notes and compiled results into a single report for each session. When the conference ended the collective wisdom from every subgroup session was made available to all the participants.

One opinion was heard again and again—"business-as-usual" could no longer be tolerated. This attitude, the professional forestry community realized, has contributed significantly to our current problems.

Nevertheless, the participants departed feeling encouraged, having reaffirmed their commitments and desire to work toward a common goal.

Programs as successful and well received as this are not put together overnight. At the instigation of NAPFSC leaders, preparations began with a Forestry Education Assessment study in late 1986.

This study and a detailed follow-up were presented in the *Journal of Forestry*, September 1989, as a seven-page feature titled, "Forestry Education and the Profession's Future." Anyone interested in forestland conservation and management should read it.

Professional forestry educators have had to scramble to keep pace with the changes of the last fifty years. Only a few schools have managed to stay ahead of the curve. In June of 1991, the Society of American Foresters compiled a comprehensive report, "1991 Forest Resources Enrollment and Degrees Survey Tables," as background information for the symposium steering committee.

This report was supplemented by four pages of exhibits and a two-page summary of statistics which help place the 1990 figures in context with enrollment trends over the past decade.

Total enrollment in all programs at accredited schools of forestry in 1990 was 16,397 and 2,985 degrees were granted. This compares with a peak enrollment in 1980 of 22,985, decreasing each year to a low in 1987 of 13,684. Since then there has been a gradual increase. The four broad categories for enrollments are Forestry, Fish and Wildlife, Wood Technology, and Recreation.

Forestry enrollment declined from 11,152 in 1980 to a low of 4,490 in 1988, increasing again to 5,983 in 1990. Degrees in forestry also peaked in 1980 at 2,732, dropped to 913 in 1989, and recovering slightly to 1,114 in 1990.

Accompanying the dramatic declines in the 1980s is a noticeable trend away from the Society of American Foresters accredited forestry curricula. It has now reached a point where graduates of an accredited program are nearly in the minority. And while forest school enrollment is once again rising, the renewed interest is not centered in professional forestry programs.

Environmental High Ground
vs. Professional Forestry

Forestland conservation is an environmental issue of the first order. The high ground for setting the environmental agenda is held by environmental activist leaders; industry leaders are somehow excluded. Unsurprisingly, our most talented young people are being attracted to the environmental high ground.

This trend is obvious at two private forestry institutions offering graduate programs. Yale University's School of Forestry & Environmental Studies class of 1991 included fifty-two graduates with a Master of Environmental Studies and only five graduates with a Master of Forestry degree. For the class of 1992, the numbers were forty-nine and three.

Duke University's former School of Forestry, as of 1991, became the School of the Environment. Historically Yale and Duke graduated professional forestry leaders, but this, it would appear, is no longer the primary mission of these two leading schools.

The trend at Duke away from Master of Forestry degrees toward Master of Environmental Management, similar to Yale's, is significant for the Forest Products Industry. This distressing trend is evidenced by Duke's employment profiles. The 1985 profile surveyed 1980-85 graduates. The list of employers included nineteen leading Forest Products Industry firms. In the 1989 profile, however, this number dropped to thirteen firms. And in the 1991 profile, only one Forest Products Industry firm was listed as an employer.

In short, the Forest Products Industry has failed to show young people that they can contribute to a healthier environment by enrolling in professional forestry programs. Thus, there is no way students can discover from their studies that there is nothing evil about improving the well-being of citizens while working in a strategic industry which utilizes renewable resources.

It is ironic that forest schools, along with forestry and foresters, have been slighted in nearly every sector at the very time that forestry, industrial and nonindustrial, has been working supply miracles *and* safeguarding the environment. This industry does not need to apologize to anyone for its overall record on environmental issues; on the contrary, the Forest Products Industry can stand proudly.

Forestry Education:
an Industry View

Returning to the Denver Symposium: the plenary session speakers represented government, the environmental community, education, and industry. The Forest Products Industry spokesman, R. Scott Wallinger, opened with these remarks: "It is clear from professional and conservation literature that there is no clear definition of what forestry is today. Until we agree on what the term means, until we agree what a forester is, we can't agree on what a forester must know, we can't agree on what a forestry school is, we can't agree on what a forestry curriculum must contain."[6]

Wallinger continued with several challenges to the participants: "First, we must now consider the bachelor of science in forestry or wildlife or range management to be the technician's degree in our profession. It is no longer any more appropriate for the true professional leader than it is for a doctor or lawyer or professor or business man or woman.

"Once we accept the notion that the bachelor's degree is a technician degree and will be followed by another degree at the professional level, we create enormous power to restructure the content of that degree in ways that are logical and helpful to students and universities."

His second point: "A broad and comprehensive, three-year doctor of forestry degree should be the prerequisite for full forest resource management leadership in forestry's second century. We can provide the basis in four years for a silvicultural or wildlife or range or agronomy or hydrology specialist, but we cannot cram the broad base for a true resource management professional into that time frame. It's time to quit pretending we can."

Wallinger emphasized this point by questioning "our currently inadequate concept of the profession that rests on a grossly low education base. That limitation is reflected in people who are scientifically trained—often in narrow specialties—but who lack the broader education in language, history, sociology, and ethics to lead the profession and the public policy debate."

"The problem is that we have no broadly recognized degree or vehicle, comparable to the MBA in business, to build on the undergraduate base and produce a person with the breadth and depth of scientific, administrative and social skills to manage the broader array of technologies and specialists required in forest resources today.

"The doctor of forestry degree has been present for years as a potential vehicle, but it is little used because universities demand the Ph. D. and the D.F. is little known."

Wallinger also stressed that the undergraduate degree can provide the foundation for absorbing the scientific work that extends beyond the undergraduate level ". . . plus the financial and integrative management skills to work with the full array of specialists effectively."

"So, I submit to you that a new comprehensive doctor of forestry degree should be number one on your list so that, in the near future, our profession will be led by men and women with the broader, deeper education needed to meet public expectations of a professional."

The third point was equally challenging: "The place of forestry in the university *must* be called into question. It may be okay for the bachelor's degree to come from a department. But the breadth of vision and the power of office to reach out to, and draw from, the full resources of a university are more likely to rest in the dean of a school or college reporting directly to the president of a university."

Earlier, Wallinger had noted in his presentation that eighteen of the accredited forest schools have departmental status while seventeen are schools and only eleven are colleges within their parent universities.

Further, "The resources to do forest research for our second century of forestry can no longer reside just within a school of forestry. . . .The *science* of modern forestry draws on the entire university: mathematics, statistics, biology and zoology, soils and genetics, biotechnology.

"A school capable of providing leadership must be empowered

to integrate these total resources, to influence their direction, to draw on their funding. This rarely occurs at the department level."

Linked to the "*science* of modern forestry," he emphasized that "the *management* of modern forestry draws likewise on the entire university. It draws on the sociology, the business and law schools, the language and journalism schools. Again, the power of the school to integrate these resources is crucial."

Summing up his third point, he predicted that "this would ultimately lead to fewer universities offering the more integrated doctor of forestry degrees, with others focusing on the undergraduate and specialist component. . . .

"A refocusing of resources would be helpful and could lead ultimately to significant economies in teaching and research. Today we have too much of the wrong kind of capacity for both, with too few of the broad professionals emerging that we need."

The Future of Forestry:
Trends and Critical Issues

Following this challenge, the symposium attendees enthusiastically began their workshops. A bit of the flavor of the final day comes through in this summary of "Workshop Session I: The Future of Forestry; Trends and Critical Issues":

Population Pressure. . .

- increased demands

- impact on finite resource

- need for alternative management strategies

- probable increase of regulation

- need for institutional changes to meet increasing/changing demands

- unknown/changing expectations

Need for Ecosystem Approach. . .

- to achieve integrated forest management

- to maintain biodiversity

- to ensure sustainability

- to maintain and restore productivity (for market and nonmarket values)

- to address societal concerns

- to maintain forest health

- integrating biological, social, and behavioral sciences

Effects of Social Values. . .

- changing more rapidly than profession can respond

- values are conflicting

- profession needs to anticipate and respond to changes

- need to balance values, benefits, and uses

- resolution of value conflicts must be political

- we are managing a rural resource for an urban society

- impact of regulation on resource management and supply

Future Resource Decision-Making. . .

- need to work within political system

- need to respond to societal concerns

- need to reestablish public trust

- need to increase public understanding of interaction and tradeoffs including resource allocation

- need to educate public policymakers

Global Scale. . .

- our resource decisions affect economies and societies in other nations

- others' resource decisions affect us
- different nations have different priorities regarding resources
- need to resolve inequities of resource allocation
- resource managers need global view

Information/Technology Explosion. . .

- need to integrate new knowledge and technology
- need to communicate both old and new information to public effectively

All in all, a fair indication of the complexity of the issues. Members of the Society of American Foresters and NAPFSC have already embraced the educational summit findings. Symposium proceedings were published by SAF in February 1992.

Clearly, our second hundred years of professional forestry requires a paradigm which differs from the historic Biltmore School of Forestry founded in 1898 by Dr. Schenck. We should nevertheless guard against ignoring the basics he stressed which are an integral part of our hard-won conservation ethic, an ethic that has served well.

There is no question that professional foresters, along with other members of the forestland conservation and management team, can join with citizen conservationists in doing a better job of educating the public about the many complex matters dealing with forestland conservation, such as the resilient and dynamic nature of our renewable resources. But first we must educate ourselves. The final chapter offers a means of doing so; sources of information covering the full spectrum of forestland conservation and management are listed.

Above all, let us not forget the wisdom of George Perkins Marsh who cautioned against looking to governments, whether at the federal or state level, for solutions to our conservation problems.

8

TREES AND GOVERNMENT

THROUGHOUT THE NINETEENTH century, federal legislators exhibited a conspicuous lack of interest in conservation. Indeed, if it had been left to Congress, there is no telling when a conservation ethic would have evolved.

Carl Schurz, interior secretary from 1877-81, was considered the first conservationist to hold cabinet rank. He was totally frustrated by the system of political patronage that prevailed in the Interior Department from its beginnings in 1849. As Henry Clepper noted in his *Crusade for Conservation*:

> At the American Forestry Congress held in Philadelphia in 1889, Schurz described the consequences of the government's inability to protect its public lands. Hundreds of sawmills, he said, manufactured lumber from timber stolen from them. Little attempt was made to control fire and that little was so feeble as to be hopeless. It was an indictment not calculated to inspire confidence in representative government.[1]

Carl Schurz (1829 - 1906)

Carl Schurz was born 2 March 1829, in Liblar, a little town near Cologne, Germany. He began his studies at the University of Bonn in

1846 and continued until, like so many others, he was forced to flee in the aftermath of the 1848 Revolution. Schurz immigrated to the United States in 1852 and, like thousands of his countrymen, settled in Wisconsin.

In 1860, Schurz was appointed minister to Spain, but resigned the very next year and joined the Union army. During the war, he commanded divisions at Fredericksburg, Chattanooga, and Gettysburg, attaining the rank of major general. He worked as a Washington, D.C., correspondent for the New York Tribune in 1865-66, was editor of the Detroit Daily Post, 1866-67, editor of the St. Louis Westliche Post, 1867-78, and editor of the New York Evening Post, 1881-83.

From 1869 to 1875, Schurz served as United States senator from Missouri and was appointed secretary of the interior department in the Rutherford B. Hayes administration, 1877-81, the first citizen of German birth to obtain a cabinet post. He was also one of the earliest high-ranking public officials advocating forestland conservation.

Schurz was an ardent foe of patronage and corruption and he vigorously enforced laws protecting federal forestlands. He was also the first cabinet officer to advocate federal forestland reserves and scientific management of forestland resources. His writings included the books Speeches published in 1865 and Henry Clay, 1887. Schurz died on 14 May 1906.

FOREST RESERVES ESTABLISHED, 1891

From its inception in 1882, the Committee on Legislation of the American Forestry Congress presented forestland conservation petitions and resolutions to both branches of Congress year after year, with little or no success, and with only scant encouragement from a few senators and representatives.

The primary purpose behind these efforts was to overturn the Department of Interior's mandate, given by Congress at the end of the Civil War, to dispose of public lands. The department had been granted neither authority nor funds to administer the public domain. Land dispositions were chaotic and the exploitation of natural resources went forward unchecked.

In September 1887, a draft of a bill to establish forestland reserves was presented at the Forestry Congress meeting in Springfield, Illinois. Largely the work of the corresponding secretary, Bernhard Fernow, it was modified and included in a memorial to the Senate and House of Representatives.

Nathaniel H. Egleston, former chief of the Forestry Division in the Department of Agriculture and chairman of the Forestry Congress' Committee on Legislation, appreciated the importance of forestland reserves and worked diligently for passage of the proposed bill. This effort of the Forestry Congress also won editorial support from several informed newspaper editors. As in previous years, however, Congress failed to act.

Nathaniel Egleston (1822 - 1912)

Nathaniel Hillyer Egleston was born 7 May 1822 in Hartford, Connecticut, graduated from Yale Divinity School, and became a prominent Congregational minister and teacher in Williamstown, Massachusetts. He developed an early interest in forestland conservation and in 1882, when the American Forestry Association merged with the American Forestry Congress, Egleston was elected a vice president of the combined organization.

In 1883, he was appointed the second chief of the Division of Forestry of the Agriculture Department. In his first annual report to Agriculture Secretary George Loring, he emphasized the need for effective management of public forestlands and the establishment of forest schools and forestland experiment stations. He evidently greeted his replacement by the more experienced Fernow with relief in 1886 and continued to work under him until Pinchot took over the division in 1898. Egleston died on 24 August 1912.

Despite disappointments, the American Forestry Congress' Committee on Legislation pressed on. In November 1889, two years after Fernow's memorial had failed to gain the necessary support in Congress, another major effort was organized. This time a group of committee members met with John W. Noble, secretary of the interior, the assistant secretary of agriculture, and a committee of the American Association for the Advancement of Science.

After this meeting, the Forestry Congress committee prepared a bill providing "for the reservation and protection of forest lands of the public domain, and to establish a commission to inquire into the condition of the said lands, and to report a plan for their permanent management." Members of the committee succeeded in getting the bill introduced, but were again frustrated when action was deferred until the next session.

This time, however, the frustration proved momentary. On 3 March 1891, "the Congress of the United States empowered the President to withdraw forest reserves from the public lands. President Harrison thereupon proclaimed nearly 13 million acres of forest reserves, and President Cleveland soon added nearly five million more."[2]

In addition to the American Forestry Association, Clepper credits a number of individuals as having aided in the passage of this "momentous legislation." He cites Egleston and personnel of the Division of Forestry and gives special recognition to Secretary Noble who, according to Fernow, "had been won over to the views for which the Division and the Association stood" and was able to exert considerable influence.[3]

Another who contributed effectively was Edward A. Bowers, assistant commissioner of the General Land Office and later secretary of the American Forestry Association. At the time, "Fernow asserted that the forest reserve law 'changed the entire land policy and all previous notions of the government's functions concerning the Public Domain.'" William B. Greeley, in his book, *Forest Policy*, described this law as "the first decisive action toward a national policy of forestry."[4]

William Greeley (1879 - 1955)

William Buckhout Greeley was born 6 September 1879, in Oswego, New York, graduated from the University of California in 1901, and earned a Master's Degree in Forestry from Yale University in 1904. He began his career in the Bureau of Forestry, Agriculture Department, in 1904, worked his way up through the ranks, and was appointed chief forester in 1920.

Along the way, he was supervisor of Sequoia National Forest (1906), regional forester of the Northern Rocky Mountain Region (1908), and assistant forester in charge of the branch of forest management in

Washington, D.C. (1911). During Greeley's tenure as chief forester, National Forest land increased from 154 million acres to 160 million.

During World War I, Greeley commanded the Forestry Division, American Expeditionary Forces, in France, attaining the rank of lieutenant colonel.

Greeley's early field experience in the western states convinced him that forestland conservation could best be addressed by means of cooperative programs conceived, financed, and implemented through federal, state, and industry organizations. With passage of the Clarke-McNary Act in 1924, these views prevailed over those of Pinchot, who throughout his career advocated federal control over private forestlands.

In May 1928, Greeley became secretary-manager of the West Coast Lumbermen's Association, serving in this capacity until his retirement in 1946. Throughout the Western states he was a major influence for advancing forestland conservation and management.

Greeley was elected president of the Society of American Foresters in 1915, elected a Fellow three years later, and in 1946, received the Sir William Schlich Memorial Medal. Greeley was a director of the American Forestry Association for thirty-four years and a frequent contributor of articles on the subject of natural resources to both technical and general interest magazines. He authored two books, Forest and Men (1951) and Forest Policy (1953).

The University of California awarded him an honorary LL.D. Degree in 1927. Greeley was founder and first president of the Yale Forest School Alumni Association and in 1955 received the Yale University Medal for Outstanding Service. Yale would also name its new forest research laboratory in his memory in 1959. Greeley died on 30 November 1955.

FOREST MANAGEMENT ACT, 1897

Unfortunately, the bill authorizing the establishment of the forestland reserves did not follow the model bill drafted by Bernhard Fernow. It failed, for example, to provide for the practical management of the reserves, protecting them from wildfire and theft. And though each year the American Forestry Association proposed corrective legislation, Congress again turned a deaf ear. For six years nothing was done.

Then in 1896, searching for influential supporters, the Forestry Association "persuaded Hoke Smith, Secretary of the Interior, to request the National Academy of Sciences to investigate and report on a rational policy for the forested lands of the United States."[5]

The academy had been created by Congress in 1863 for the purpose of serving as an advisor to the federal government in matters of science and technology. Its president, Wolcott Gibbs, assigned a forestry committee to the task, which ultimately gave the proposed legislation the needed boost.

"Charles Sprague Sargent, head of Arnold Arboretum of Harvard University, an eminent botanist, was named chairman of the National Academy of Sciences' committee. Gifford Pinchot, although not a member of the Academy, was appointed to the committee as secretary."[6] Other members included Alexander Agassiz of Harvard, army engineer Henry L. Abbott, William H. Brewer of Yale, and geologist Arnold Hague.

The committee's assignment seems monumental in scope. It was given a year to make a study of public lands, which at the time comprised some 600 million acres, and to make legislative proposals for the forestland reserves, then totaling nearly 20 million acres.

The committee's report was presented to the secretary of the interior on 1 May 1897. Mincing no words, it declared that "Interior's methods and personnel for handling the reserves were inadequate. It noted that the Department's politically appointed employees were unable to cope with such basic needs as protecting the lands from fire and theft."[7]

The committee's recommendations for administration of federal forestlands, though "wholly rational and practical," initially seemed to have little effect. On 4 June 1897, however, Congress did attach, as a rider to the civil appropriations bill, the so-called Forest Management Act.

Fernow was less than pleased with the new law, describing it as imperfect. But Greeley disagreed, contending, "It was a well-drawn law which has remained the basis of operating the National Forests to the present time. One of its significant phrases was the declaration that the reserves were created 'for the purpose of securing favorable conditions of water flows and to furnish a continuous supply of timber for the use and necessities of citizens of the United States.'"

This, Greeley observed, constituted "the first clear, realistic step in a public policy of conserving rather than distributing forest resources. . . ."[8]

The Forest Management Act did indeed establish long-sought regulations for occupancy and use of federal forestlands as well as their protection from wildfire and theft. Thus, after fifteen years, a small but persistent band of concerned citizens was able to convince a reluctant Congress to authorize the creation of forestland reserves (1891) and provide for their protection and administration (1897). Another distinct step had been taken toward developing a conservation ethic for trees and forestlands.

FORESTLAND CONSERVATION
AT THE STATE LEVEL

At the end of the nineteenth century, the American Forestry Association membership could look back on its first twenty-five years with a degree of satisfaction. This small group of dedicated conservationists had influenced federal legislation at a time when their proposed laws were anything but popular.

They were also active at the state level. Furthermore, they were increasing their influence on the American public and stimulating interest in forestland conservation, partly through the official publication, *The Forester*, which was circulated not only to some 1,500 members, but also to two hundred libraries and nonmember subscribers. Its editorial bent, then as now, was mainly nontechnical, in keeping with its goal of rousing public interest. Magazine editors also supported many projects which had not originated with the Forestry Association itself, Arbor Day being a prime example. Clepper notes that as early as 1883, "a special committee was appointed to advance the movement for observance of Arbor Day, particularly in the schools. . . .Within a few years, governors were proclaiming Arbor Days in schools and communities were celebrating it as a holiday."[9]

J. Sterling Morton (1832 - 1902)

Julius Sterling Morton, the originator of Arbor Day, was born 22 April 1832 in Adams, New York. He graduated from the University of Michigan in 1854 and took up a homestead that same year in the Nebraska Territory, where he edited a local newspaper. Morton was elected to the Territorial Assembly in 1855 and in 1858 was appointed secretary of the Territory by President Buchanan. At times during this period, he served as acting governor and as a member on the Board of Agriculture.

Morton is best remembered for his 1872 proposal to the Nebraska Board of Agriculture that a day be set aside for tree planting. The state legislature officially designated 22 April (Morton's birthday) as Arbor Day, a legal holiday. Other states would follow Nebraska's example, and today Arbor Day is observed in every state and in many nations the world over.

Morton was also influential in inducing Congress to pass the Timber Culture Act of 1873. This law, until its repeal in 1891, offered free land to settlers who would plant trees on their claims. In 1893 Morton was appointed secretary of agriculture by President Cleveland. That same year he was elected the ninth president of the American Forestry Association, serving four terms. Morton died 27 April 1902 in Lake Forest, Illinois.

The Forestry Association's accomplishments at the state level were impressive. It played a leading role in encouraging citizens to establish and support state forestry associations and worked diligently for the creation of state forestry boards, commissions, and departments.

The first state forestry association was organized in Minnesota in 1876, and from the beginning, there was a close working relationship between it and the national organization. The establishment of three other state forestry associations followed in 1885—Colorado, New York, and Ohio.

The Pennsylvania Forestry Association, the oldest in the nation with an unbroken record of activity, was founded in 1886 under the leadership and presidency of Dr. Joseph Rothrock. Illinois and Texas followed in 1888.

Wisconsin was next in 1893, followed in 1894 by the New Jersey Forestry Association, which became "an early and constant cooperator of the American Forestry Association. A joint meeting of the two groups was held in south Jersey in May 1895."[10] American Forestry Association leaders were so taken by *The Forester*, the official publication of the New Jersey group, that by 1898 they were able to adopt it as their own.

The Connecticut Forestry Association was organized in 1895, joined in 1897 by North Carolina, Oregon, and Washington, and by Massachusetts in 1898.

Today some thirty-five states have active forestry associations representing their members' interests in forestland regulations, tax and insurance rates, environmental regulations, and other state legislative issues. Many have ongoing forestry education programs at the grass roots level.

Nearly all of the state association leaders are active in the National Council of Forestry Association Executives, which meets annually.

In addition to these citizens' state forestry associations, by the close of the nineteenth century twelve states had established boards, commissions, or departments of forestry to pursue forestland conservation and management as authorized by their respective legislatures. Today, thanks to the tremendous success of cooperative forestry efforts at all levels of government and throughout the private sector, there are state forestry agencies in all fifty states as well as in Guam, Puerto Rico, and the Virgin Islands.

These agencies come together under the umbrella of the National Association of State Foresters, with an office in Washington, D.C. The cooperative programs of the state agencies and the Forest Service continue to have a positive impact on forestland stewardship with educational programs at all levels.

Meanwhile, to the north, the Canadian Forestry Association was organized in Ottawa on 8 March 1900. A number of its officers had been active in the American Forestry Association and it patterned its programs for conservation and management after its counterpart in the United States. Today this forestry association comprises nine federated provincial groups with twelve thousand members.

AMERICAN FOREST CONGRESS, 1905

Progress in forestland conservation captured the nation's attention in early January 1905 when delegates to the Forest Congress were greeted at the White House by President Theodore Roosevelt. Sponsored by the American Forestry Association, this congress gave a significant boost to the budding conservation ethic for trees and forestlands.

Clepper notes its objectives: "to establish a broader understanding of the forest and its relation to the great industries depending upon it; to advance conservative use of forest resources for both the present and future needs of these industries; and to stimulate and unite all efforts to perpetuate the forest as a permanent resource of the Nation."[11]

Each conference session was convened by James Wilson, secretary of agriculture and president of the American Forestry Association. President Roosevelt's address, entitled "The Forest in the Life of the Nation," was delivered on 5 January in the National Theater to an audience numbering two thousand.

The Forest Congress brought to a climax several initiatives which had been endorsed by the association for a number of years, the most significant of which favored bringing all forestry work by the federal government under the aegis of the Department of Agriculture. Up to that time, forestry efforts at the federal level had been distributed among three agencies; the General Land Office, the Geological Survey in the Department of the Interior, and the Division of Forestry in the Department of Agriculture.

The Forestry Association and James Wilson were not alone in supporting this change. Both Interior Secretary E. A. Hitchcock and Land Office Commissioner William A. Richards endorsed the move. Finally, and most importantly, so too did President Roosevelt.

The deliberations of the Forest Congress were wide ranging. In all, eighteen resolutions were adopted by the delegates. Practically all of these, in whole or in part, were subsequently acted upon.

One called for the expansion of professional forestry educational opportunities in colleges and universities and an increase in the general awareness of conservation at the secondary school level. Another urged that states, industry, and the public at large protect forestlands from wildfires. Still another sought to augment research in forestry, especially by agricultural experiment stations.

A forerunner of the development of a national park system was a resolution "calling on the federal government to purchase the Calaveras Grove of Big Trees in California and the reconveyance by the State of California to the federal government of Yosemite Park for administration as a national park."[12]

Within the states, one resolution called for the organization of forestry departments to advance reforestation efforts. Another resolution urged the establishment of forestland reserves in the Southern Appalachian Mountains and in the White Mountains of New Hampshire.

While the number of registered delegates to the Forest Congress totaled a modest 385, the average attendance at the eight sessions was more than one thousand—an "impressive assemblage of concerned citizens from all spheres of society."[13]

In addition to civic leaders, editors, educators, federal and state legislators, and the elite of the growing conservation movement, a number of representatives from industry chose to come forward and assume a public responsibility. Mining, railroad, and Forest Products Industry companies were represented along with farming and livestock interests.

The Forest Congress was truly a turning point in the conservation movement, for much of what both Chief Forester Pinchot and President Roosevelt were seeking came to pass. When Pinchot succeeded Fernow as chief forester back in 1898, he inherited what was considered to be the only staff with expertise in forestland conservation and management at the federal level. The forestland reserves and their administration, however, remained in the Department of the Interior.

President Roosevelt understood Pinchot's problem and in his first message to Congress in 1901, he urged the transfer of the forestland reserves to Agriculture; but Congress dallied for four more years. Thirty days after the 1905 Forest Congress, however, the administration of the reserves was assigned to the Department of Agriculture.

The Bureau of Forestry became the Forest Service by a separate act of Congress on 3 March 1905 and the name Forestland Reserves was changed to National Forests two year later. Pinchot became the first chief of the Forest Service as we know it today, though he was not the first Chief of Forestry in the Department of Agriculture.

The Agriculture Department had been established in 1862 and obtained cabinet rank in 1889. As mentioned earlier, Dr. Hough was appointed the first federal forestry agent in 1876, and when the Division of Forestry was established in 1881 (renamed Bureau of Forestry in 1901), he became its first chief.

Egleston succeeded Dr. Hough in 1883, followed in 1886 by Fernow and in 1898 by Pinchot. Technically, according to the *Encyclopedia of American Forest and Conservation History*, Pinchot was not a "chief." The head forestry administrator was called agent from 1876 to 1880; chief from 1881 to 1898; forester from 1898 to 1935, and chief again from 1935 to the present.

Pinchot himself changed the title from chief to forester in 1898. Dana observed in his *Forest and Range Policy*, "'In Washington chiefs of division were as thick as leaves in Vallombrosa,' according to his [Pinchot's] view but there was only one Forester spelled with a capital letter."[14]

Confusion over titles and the early management is thus to be expected. But it fails to excuse those who write authoritatively and erroneously about a subject they have studied little, if at all. A case in point is an article, "American Survey, The Forest Service," which appeared in the 10 March 1990 issue of *The Economist*.

It states: "The agency [Forest Service] had heroic beginnings, founded by Gifford Pinchot, a forester, to save wild lands from the depredations of the robber barons."[15] With such a beginning, only the uninformed would bother reading further.

CONFERENCE OF GOVERNORS, 1908

From the very first day, Pinchot approached his job with bursting enthusiasm and soon was recognized as the nation's leading crusader for forestland conservation. During his twelve-year tenure as Forester, National Forest land increased dramatically from 40 million acres to 172 million. Henry Graves, who succeeded Pinchot in 1910, best described the daunting task of the Forest Service. Speaking of the National Forest holdings, Graves observed:

> This was wildland, much of it in frontier country that had never been surveyed or, from the standpoint of character and condition of the forests, accurately described. It was an immense task to organize the management of these areas. It involved problems of determination of boundaries; land surveys; construction of roads, trails, and telephone lines; construction of ranger stations and other buildings; exploration and description of the forests; development of specific policies and procedures for disposal of timber; control of use of forage for livestock, and a multitude of other features of forest utilization; problems of building an adequate and competent administrative personnel; initiation of organized fire control; and development of techniques in applied forestry, range management, and protection of wildlife.[16]

Pinchot had the good fortune to have a staunch ally in the person of President Roosevelt, who also energetically crusaded for conservation. Inspired in part by the success of the American Forest Congress in 1905, they turned their attention to the conservation of all resources and together set an agenda for the first national conference on conservation. It was held in the East Room of the White House 13-15 May 1908 and was officially designated the Conference of Governors. This unprecedented undertaking attracted worldwide attention.

Conferees included members of the cabinet, the Supreme Court, and a good many congressmen. Also present were fifty-two governors of states, territories, and dependencies and representatives of seventy associations, societies, and labor organizations. There were twenty-one newspaper reporters and fifty-two special and general guests as well as members of the Inland Waterways Commission.

The conference agenda covered the gamut of mineral, land, and water resource concerns. Although the subject of wildlife was overlooked and pollution was as yet not considered important, conservation was reverently discussed session after session.

President Roosevelt gave the opening address. Other speakers included Andrew Carnegie, the steel magnate, who began the second session with an address on the conservation of ores and related minerals, and James J. Hill, railroad entrepreneur, who opened the third session with a discussion on the natural wealth of the land.

Hill's speech was followed by a keynote address, "Forest Conservation," by Robert A. Long, president of the Long-Bell Lumber Company. This was no new role for Long. In 1903, as president of the Southern Lumber Manufacturers' Association, he had "sounded a call for conservation," warning of wasteful and rapid depleting of forestland resources if current lumber practices continued. Now he advocated the conservation of forestland resources as "practical and necessary for the healthy growth of the industry."[17]

According to Clepper:

> The information provided in the speeches and discussions opened the conference delegates' eyes to conditions and opportunities to which many had previously been blind. Most conferees, aware that they were participating in a historic event, rose to the occasion by putting politics aside in order to present a united front. Impressed by the importance of resource conservation, they felt impelled to help advance it.[18]

The conference recommended that each state form a commission on natural resources. Acting for the federal government, President Roosevelt appointed a National Conservation Commission on 8 June 1908. It was organized in four sections: land, water, forestland, and mineral resources, each one chaired by a member of Congress.

The commission had forty-eight members and Pinchot was elected commission chairman. The president directed the commission to investigate the condition of the country's natural resources, advising him of their findings. Again, the scope of organization and participation was unprecedented.

JOINT CONSERVATION CONFERENCE, 1908

Although the commission was without funds, Pinchot was able to carry out the assignment; on 1 December, inventory summaries of the four sections were accepted and combined into a report that became the basis for the Joint Conservation Conference held a week later in Washington, D.C.

The nearly 500 conference attendees included 22 state governors, 11 personal representatives of governors, 98 representatives of 31 state conservation commissions, 105 representatives of 59 scientific and labor organizations, 38 chiefs and experts of 16 federal bureaus, and 68 delegates at large.

Clepper states that, on balance, those attending the Joint Conservation Conference were "better informed and generally more knowledgeable about resources than were the participants in the White House Conference held in May."[19]

Among foresters in attendance were Henry Hardtner of Louisiana, Herman von Schrenk of Missouri, Philip Ayres of New Hampshire, Alfred Gaskill of New Jersey, and Dr. Joseph Rothrock of Pennsylvania. Representing the Forest Service were William Cox, William Hall, Royal Kellogg, and Edwin Ziegler, all of whom became prominent in the profession.

This conference also addressed the subject of wildlife, which had been omitted from the May conference. "The oversight was corrected in the report of the National Conservation Commission, which contained a comprehensive statement, 'Relations of Birds and Mammals to the National Resources,' by C. Hart Merriam, Chief of the Biological Survey."[20]

After presenting their findings, the conferees urged that conservation commissions be maintained in all the states and that Congress maintain a national commission on the conservation of resources.

President Roosevelt, in a special message to Congress on 22 January 1909, transmitted the commission's report and requested appropriation of $50,000 to cover the commission's expenses.

But Roosevelt was at the end of his term, and Congress not only declined to appropriate the funds, it prohibited all bureaus from doing work for any presidential commission, board, or similar body appointed without congressional approval. Progress along these lines came to an abrupt stop.

WEEKS LAW, 1911

Meanwhile, the scourge of uncontrolled wildfires was burning its way into the consciousness of people from all walks of life. A consensus opinion was forming to the effect that wildfire was the number one issue in forestland conservation.

Pinchot was a notable exception. He was convinced that the Forest Products Industry was mainly culpable for forestland devastation, even though wildfires consumed a greater volume each year than was harvested by logging. This view notwithstanding, P i n - chot encouraged his regional forester, Greeley, to build cooperative wildfire control organizations in Idaho and Montana. The wildfires in the Far West in 1902 had already convinced industry leaders of the need for cooperation. Major lumber producers, led by George S. Long of the Weyerhaeuser Timber Company (not related to Robert A. Long), joined forces to establish protective associations in the Pacific Northwest.

This group invited Greeley to a conference in Spokane, Washington, on 4 January 1909, to draft uniform wildfire laws in Oregon, Washington, Idaho, and Montana. During this meeting, Greeley pushed for the completion of binding cooperative agreements, and drafts were forwarded to Chief Forester Pinchot.

The policy, as laid out by Greeley with two additions offered by the lumbermen, became the standard form of agreements with the northern Idaho forestland protective associations:

1. The district Forester and representatives of the Associations would establish protective districts based on areas owned by each and agree upon a prorated division of expenses for *large* fires.
2. All expenditures for *large* fires would be recorded on Service vouchers and receipt forms.

3. All Association wardens would be appointed forest guards at a nominal salary.
4. Supervisors and local Association representatives would arrange practical division of fire patrols.
5. Proper division of *normal* fire fighting expenses would be determined by a forest supervisor, or by a warden or ranger operating under his instructions.

Leaders of the lumbermen added provisions that:
1. One man, either a ranger or warden, would have charge of fire fighting in each district.
2. Local representatives of the Service and Association would agree in advance to a wage scale for temporary laborers.

The Chief Forester approved this standard policy in its entirety—including the prorating of expenses for extinguishing large fires.

> At a joint meeting of the northern Idaho protective groups held in Spokane on May 6, 1909, the delegates voted unanimous approval and the general agreement was executed 'on the spot' by Greeley and the lumbermen.[21]

Greeley's only remaining task was to conclude final arrangements with individual industry associations. Within a month, cooperative districts were established with the Pend Oreille Timber Protective Association, the Coeur d'Alene National Forest, the Potlatch Timber Protective Association, the Coeur d'Alene Protective Association, and the Clearwater Timber Protective Association.

Following these agreements with industry, Greeley successfully concluded negotiations with the Spokane-Inland Empire Railway, Northern Pacific, and the Great Northern Railway. At this point, the largest private land owners were in the cooperative camp; the only remaining problem concerned the state lands, which were without an effective protective system of their own.

As a solution, Greeley proposed to the governor of Montana

and the Idaho Board of Land Commissioners that forest rangers be appointed state fire wardens until the state legislatures provided for their own forestry programs. In less than two weeks, the Montana Assembly passed a forestry bill incorporating Greeley's plan. Similar action in Idaho came at a later date.

The concern for the protection of forestlands from wildfire culminated in passage of the Weeks bill in 1911; it had been shaping up for a number of years. Three presidents had endorsed it. The Senate had voted favorably three times and the House once. Moving with glacial speed, it finally received President Taft's signature on 1 March.

The Weeks Law established a national forestry policy and provided for its implementation. Congress authorized cooperation between states for both forestland conservation and watershed protection, for the first time subsidizing effective wildfire detection and control measures. An appropriation of $200,000 was provided for the secretary of agriculture to implement these programs.

Congress also appropriated $1 million for the fiscal year ending 30 June 1911 and up to $2 million for each of the next four years for survey and acquisition of forestlands in the headwaters of navigable streams. There was no geographic restriction for this appropriation which allowed purchases of land for the National Forest System in whatever location state legislatures approved.

"Until the law was amended by the Clarke-McNary Act of 1924, National Forest purchases aggregated 2.5 million acres at a total cost of $12.5 million."[22]

The Weeks Law had far-reaching implications for cooperative wildfire detection and control programs. Eventually, every state adopted legislation to protect forestlands from wildfires, as only those states which had by law organized such systems were eligible for federal funds:

> In 1911, the year the law was enacted, only 11 states were doing sufficient work in forest protection to qualify for the financial assistance available from the federal government. The acreage under protection was 60 million acres. In these 11 states, the federal contribution was $36,692; state contributions amounted to

$165,975; and private interests put up $54,590. When, in 1924, the Weeks Law was amended by the Clarke-McNary Act, 29 states were cooperating. The protected area had risen to 178 million acres. Federal seed money that year was $397,357; state, $1,662,532; and private, $1,850,862.[23]

These figures are most interesting in view of the bad reputation the industry had acquired with its history of cut-out-and-get-out. The figures evidence the impact that the Weeks Law had on forestland conservation in the private sector as well as on state legislatures.

Both the number of states participating and the acreage protected tripled during the thirteen years the law was in effect. Federal seed money increased elevenfold during the period, while state funding increased tenfold. But the real change came from the private sector where, with minimal encouragement by government agencies, funding for cooperative forestland conservation increased by a factor of thirty-four.

COOPERATIVE FORESTRY

With the signing of the Weeks Law and on the strength of Greeley's success with cooperative forestry programs in Idaho and Montana, Chief Forester Graves, who had replaced Pinchot in 1910, invited Greeley to Washington, D.C., to carry on his efforts throughout the National Forest System.

From Washington, Greeley continued to explore the merits of cooperative forestry and to act upon his convictions. He knew, of course, that the Weeks Law placed a top priority on the protection of forestlands from wildfire. His vision of forestland conservation and cooperative forestry, however, went well beyond this.

The support Greeley received from industry leaders in the Western states was significant. In January 1909, when he met with forestland owners to work out cooperative forestland protection agreements, this same group laid the groundwork for what was to become the Western Forestry and Conservation Association, which throughout its history has actively supported broad-based cooperative forestry programs.

Greeley's field work in the forested regions of the United States convinced him that lumbermen had constructive ideas other than protecting forestlands from wildfire and that it was up to the Forest Service to determine how best to incorporate these ideas.

COOPERATIVE FORESTRY: STUMBLING BLOCKS

Even before 1909, leaders in the private sector were voicing their concern that business constraints beyond their control prevented them from practicing forestland conservation. For example, as cited earlier, Robert A. Long called for conservation at the 1903 meeting of the Southern Lumber Manufacturers' Association. After warning of wasteful and rapid depletion of stumpage, he prophesied:

"Within ten years large harvests of timber in the Lake States would be a thing of the past, supply of southern yellow pine would give out in eighteen years, east Texas pineries would be exhausted in twenty-five years, and all timber producing areas would be severely diminished by 1945."[24]   A pretty accurate prophesy at that time.

In Long's address before the Conference of Governors in 1908, he made the point that taxes on forestland did not encourage conservation or allow for reforestation "because timber crops were taxed annually until depleted, and cutover land was taxed practically at full value."

He went on: "The crop of the farmer is taxed when it is ready for the market, and no crop is taxed more than once. . . .The farmer's crop matures yearly, the timber owner's once in about a hundred years."[25]

Three years after Greeley's arrival in Washington, D.C., he finally was able to investigate the plight of lumbermen. This came about in 1914, when "the Forest Service, in cooperation with the Bureau of Corporations and the Federal Trade Commission, conducted a two-year study of conditions in the badly depressed lumber industry."[26]

The aim of this investigation was to "obtain and place before the public in a constructive way the essential facts regarding this industry and their bearing upon forest conservation."[27]

For Greeley the opportunity was timely. He had long wondered why lumbermen seemed determined to bring about their own demise. If there were good businessmen among them—and there had to be—surely they realized that their present methods could not long continue.

Why weren't they accepting of the principles of forestland stewardship? Could it be ascribed to mere obstinacy? Greeley couldn't believe that. His 1916 report, "Some Public and Economic Aspects of the Lumber Industry," summarized his findings.

The report "fell like a bombshell into the midst of the ardent Pinchot-led conservationists."[28] Pinchot, remember, was convinced that without exception, the Forest Products Industry was bad to the core when it came to exploiting forestlands, a conviction he carried to his grave.

"Amidst fervent protestations of shock, indignation and incredulity Pinchot labeled the study 'one of the ablest I have ever seen, and altogether the most dangerous.' In his opinion it accepted the 'commercial demands of the lumber industry as supreme over the need of forest conservation and the rights of the public.'"[29]

Pinchot evidently confused "commercial demands" with consumer demands and the rights of the public to consume. He never asked the question, "Are the producers greedy or the people who demand their products and services?"

While Greeley was addressing the problems of the lumber industry at the federal level, his report provided industry leaders with an unusual opportunity to address the public's beliefs concerning the industry's seeming unwillingness to pursue a reasonable forestland conservation policy.

Industry leaders had to admit to their own history and little about it was pretty. Over the years, they had become defensive, even reclusive. Dialogue had largely been avoided.

It had become popular to charge the industry with irresponsibility; Pinchot was not the exception, he was only the most prominent of the detractors. In the process, citizens seemed somehow able to separate their daily forestland needs from the harvesting of raw material from these same forestlands. It is not a recent phenomenon.

Some characteristics of the industry have changed little if at all over the years. It has always been a fragmented, risky business with a high attrition rate. The survivors are noted for being hardheaded, independent, and as competitive as jungle fighters.

It was easy for the general public to blame them for exploiting the nation's forestland resources. It was easy in part because it was true. Moreover, there was an aesthetic side to it all, and the ugly consequences of cut-out-and-get-out were also laid at the feet of industry.

Little thought was given to the industry's economic problems— that is, until the government's two-year study. This landmark study suggested that it was impossible for lumbermen to manage forestlands with a view to the future; they had no choice but to migrate from one forested region to another.

Fueling the reputation of those in the industry as exploiters were the many stories of skulduggery and outright theft in obtaining access to raw material on public lands. As the industry moved from one area to another, it was frequently preceded by land speculators who were adept at gaining control of vast acreages of forestland.

In some cases, lumbermen doubtless cooperated with these speculators and in many more instances they had no choice. But whether raw material was gained legally or otherwise, the fact remained that what seemed like a "low cost" investment, a steal, was seldom that, particularly when working on borrowed money. The high interest rates became more burdensome as operators borrowed additional money for mill construction and working capital.

On top of the interest burden, a number of local governments began to tax cutover land at nearly the same rate as forested land. And finally, in both good and bad markets, competition was fierce. As a result, only the highest quality logs were transported to the mills, and only the highest quality and highest value products were shipped to market. Waste in the woods, in the mills and in the marketplace was mind-boggling.

Added up, these factors left industry leaders no choice; they were compelled to cut-out-and-get-out as fast as possible. Facing uninsurable risk from wildfire and uncertain tax rates, they simply could not conceive of holding cutover land and assuming the unknown costs of growing a new forest. A few commented that the federal government could afford this but not private owners.

Nonetheless, a few operators, most notably in the Lake States, succeeded in putting aside capital and were able to reinvest in forest-lands in the Southern pine region and the Far West.

After the Weeks Law was enacted, these same operators active-ly supported cooperative forestry programs to protect forestland from wildfire; but reforestation remained beyond their means. After the turn of the century, however, and on reaching the Pacific Coast, more and more operators began to realize that their future raw mate-rial supply was in serious jeopardy.

Greeley's report made it clear that the concept of cooperative forestry needed to be extended beyond the protection of forestlands from wildfire and into such areas as forestland taxation, insurance, reforestation, and research. But just as he was arriving at nationwide cooperative forestry, World War I intervened.

Greeley arrived in France on 21 August 1917 and for the next twenty-three months played a major role in supplying wood products to the American Expeditionary Force. As commanding officer of the Forestry Division, 20th Engineers, his skills were sharpened as he fos-tered cooperation between the military and the conservation-mind-ed French forestland owners.

Greeley was responsible for the operation of ninety-five sawmills. In his commendation for the Distinguished Service Medal, General Edgar Jadwin said of Greeley, "By his engineering and exec-utive ability and tact of the highest order, he provided the supplies of timber needed by the American Expeditionary Force." The Forestry Division supplied some 300 million board feet of lumber and ties, nearly 3 million poles and piles of all sizes, and nearly a third of a mil-lion cords of fuelwood.

COOPERATIVE FORESTRY VS. FEDERAL REGULATION

Almost immediately upon Greeley's return from France, Chief Forester Graves announced his retirement and Greeley succeeded him. About that time, Pinchot assumed leadership of the Society of American Foresters Committee for the Application of Forestry.

In November 1919, Pinchot issued a report which embodied his policy of strict federal regulation of privately owned forestlands. The report opened with the statement that "within less than fifty years, our present timber shortage will have become a blighting timber famine" and went on to describe how once-productive forestlands had been "transformed by . . . lumbering into non-productive wastes of blackened stumps and bleaching snags."[30]

Pinchot would continue to seek support for his views by preaching "timber famine" and "monopoly" by a small group of operators in the Far West. His report argued that "the evidence of the past thirty years or more [has] made some basic facts abundantly clear:

1. The United States is the world's greatest timber consumer.
2. The bulk of all our standing timber is privately owned.
3. The privately owned forests have been and are being devastated.
4. The acreage of idle forest lands is already enormous and is rapidly increasing.
5. A timber shortage has already developed.
6. The timber shortage will soon become more acute.
7. The timber shortage is due to timber devastation.
8. Nothing yet done or heretofore proposed offers an adequate remedy.
9. The only possible remedy is to keep enough forest land growing trees.
10. To maintain our forests in continuous production is easily practicable."

Pinchot further clarified his position in a December 1919 *Journal of Forestry* article, "The Lines Are Drawn," in which he declared a "fight has now begun. . . I use the word fight because I mean precisely that." Then he threw down the gauntlet to the nation's foresters—"He who is not for forestry is against it."

Actually few foresters, including Greeley, could take exception to Pinchot's contention. But they could and did differ with his proposed solution. Greeley, for one, was not in favor of direct federal police action in the nation's forestlands.

Greeley's style, by now habitual, emphasized a cooperative

approach. Put another way, he favored offering a carrot to the private sector while Pinchot preferred a stick, in this instance a powerful federal stick with which to beat the Forest Products Industry into submission; and in this assault, he would hold no constitutional qualms.

This is not to suggest that Greeley opposed federal involvement; he didn't. Indeed, he supported enabling legislative efforts in five areas:

1. Federal cooperation in fire protection and forest renewal. (This was merely an amendment to and a means of continuing the forest conservation programs made possible by the Weeks Law.)
2. The extension and consolidation of federal forest holdings. (A further extension of the Weeks Law.)
3. Reforestation of denuded federal lands.
4. A study of forest taxation and insurance.
5. The survey and classification of forest resources. (Under this point he urged Congress to halt the reduction of appropriations for forest products research and to maintain and increase the number of experiment stations.)

Greeley also urged state legislatures to participate by providing for protection of forestland from wildfire and for reforestation on privately owned lands and by revising tax policies on these lands. In addition, he advocated increasing state, county, and municipal forestland ownership.

With this program, he was certain reforestation would progress further "by beginning at the bottom instead of at the top" and that individual responsibility of the use of forestlands would "actually get more tangible results, more forest growth, by working it out State by State or section by section through . . . local agencies."[31]

Pinchot's position was placed before Congress in May 1920 by Senator Arthur Capper of Kansas. Popularly known as the Capper Report, the bill sought "to prevent the devastation of forest lands." It would never reach the Senate floor for a vote.

Greeley's response was drafted and introduced before the House of Representatives by Bertrand H. Snell. The Snell Bill (ridiculed by

Pinchot as the "smell" bill) managed to reach the House floor but failed to obtain the necessary votes.

CLARKE-MCNARY ACT, 1924

The confrontation between cooperative forestry initiatives and federal regulation of private forestland continued unabated into early 1923, when the Senate adopted a resolution to establish a committee to "investigate problems relating to reforestation, with a view to establishing a comprehensive national policy. . . in order to insure a perpetual supply of timber for the use and necessities of citizens of the United States." [32]

After a series of twenty-four hearings held throughout the nation's major forestland regions, the report of this Senate Select Committee on Reforestation confirmed much of what Greeley had been advocating in his many appearances before congressional committees.

On 15 December 1923, Senator Charles L. McNary of Oregon introduced a bill containing the committee's recommendation, and a month later, Representative John D. Clarke of New York introduced the companion bill in the House. On the last day in the 68th Congress, 6 June 1924, the Clarke-McNary bill passed both houses and was signed into law by President Coolidge.

The Clarke-McNary Act established fundamental forestland policy and provided for sweeping cooperative programs in wildfire detection and control, reforestation and farm forestry extension which would remain effective into the second half of the twentieth century. To review the history of this enabling legislation is to realize that many people besides Pinchot and Chief Forester Greeley were involved.

Members of the Society of American Foresters were polled early on, and they favored Pinchot's position for federal regulation; but this polling offered no choice between federal and state regulation. Given the choice, support for Pinchot disappeared and the majority favored cooperation between federal, state, and private interests.

Greeley's position was strengthened by the solid backing of Dean Graves at Yale. Graves was then also president of the American Forestry Association, long a supporter of cooperative efforts. Greeley and Graves had been colleagues since their years at Yale and together they made a formidable team.

Clepper, in *Crusade for Conservation*, contends that Graves was the most knowledgeable president on forestry matters in the American Forestry Association's forty-nine-year history. His views were comprehensive, including wildlife consideration, scenic preservation, watershed protection, parks and recreation, as well as forestland management.

Graves had been elected president of the association following the resignation of Charles Lathrop Pack in 1922. In one of his first messages to members, he announced a three-part mission for the association: first, to become a great educational institution; second, to marshall the resources of the existing educational institutions in support of the forestry movement; and third, to assume leadership in formulating forestry policy.

To help solve the problem of forestland conservation, which Graves considered gigantic, he offered three proposals:

One, he urged Congress to appropriate $100 million over a ten-year period to acquire additional National Forest land, not only to protect watersheds but also to provide forest products. He also recommended that these lands be used as demonstration areas for the private sector as examples of enlightened protection of forestlands from wildfire.

Two, he pushed for cooperation between federal and state agencies for wildfire detection and control on both public and private lands. He again stressed the importance of research and experimentation in reforestation and the need for public education.

Three, he proposed that all lands suitable for forests in the unreserved public domain be included in the National Forest system.

Graves' proposals were timed to coincide with the work of the Senate Select Committee on Reforestation. These coordinated efforts proved decisive in carrying the day for cooperative forestry and conservation.

When the Clarke-McNary Act was finally signed into law, it included Greeley's five points:

"Sections 1 and 2 established cooperative fire prevention; section

3 authorized an extensive study of tax policies to aid the states in devising laws designed to encourage conservation and forest planting; section 4 allotted funds for cooperative reforestation of denuded lands; and sections 5, 6, and 7 authorized the extension of national forests."

Almost immediately, federal funds were expended, matching state allocations on a 50-50 basis, for growing nursery stock and planting shelterbelts, windbreaks, and reforesting cutover lands. On the same basis, funds were provided for educational assistance and technical advice to farm woodland owners.

The bill eased restrictions contained in the Weeks Law and permitted acquisition of land suitable for commercial forestlands as well as lands designated for watershed protection. Both the Weeks Law and the Clarke-McNary Act were enhanced in 1928 with the passage of the McNary-Woodruff Act, authorizing additional funds for the purchase of forestlands.

MCSWEENEY-MCNARY FOREST RESEARCH ACT, 1928

This act established a ten-year program of forestry research and included a forestlands survey. Actually, research had begun with Fernow, the first professional forester in government service, when he was named chief of the Forestry Division in 1886. He introduced the study of wood utilization, timber physics, and other related subjects that resulted in "an impressive series of technical bulletins and monographs."

Greeley's book, *Forests and Men*, described the impact of the McSweeney-McNary Act:

> It set up a permanent plan of forest research with authorization for continuing funds far beyond what the Forest Service had been able to wrangle in its current supply bills.
>
> The resources of the Forest Products Laboratory were greatly increased. [This lab was formally opened on 4 June 1910 in Madison, Wisconsin.] A series of regional Forest Experiment Stations were provided, where the tree-growing techniques and forest economics of the

local woodlands could be worked out for the benefit of all owners.

"Co-operation" was written in every paragraph of this constructive law. In almost every instance, the tree-growing stations located study tracts on private forests. Industry men were put on advisory committees, proposed study projects, and helped in the preparation of reports.

Soon most of the experiment stations were demonstration areas of local cutting and planting methods. They became meeting grounds for farmers and lumbermen. Here was more education and co-operation of a practical kind.[33]

Today the Forest Service has Research Station headquarters in Ogden, Utah; St. Paul, Minnesota; Radnor, Pennsylvania; Portland, Oregon; Berkeley, California; Fort Collins, Colorado; Asheville, North Carolina; and New Orleans, Louisiana.

The Forest Service carries out the most extensive program of integrated forestland research anywhere. As outlined in the *Report of the Forest Service FY 1992*, annual research appropriations approached $181 million directed toward scientific and technical knowledge to enhance and protect environmental quality and productivity on 1.6 billion acres of forestlands and associated rangelands in the United States.[34]

Approximately 17 percent of the annual research budget ($30 million) supported cooperative studies with forest schools, industry, and other domestic and international organizations. More than 2,800 studies are in progress at any one time.

FORESTLAND CONSERVATION: THE FIRST FIFTY YEARS

Summarizing the first fifty years of forestland conservation, the most significant accomplishments at the federal level were relatively few, and each required a great deal of perseverance.

- Forest Reserve Act, 1891

- Forest Management Act, 1897

- Transfer of Forest Reserves from Interior to Agriculture, 1905

- Weeks Law, 1911

- Clarke-McNary Act, 1924

- McSweeney-McNary Forest Research Act, 1928

But this enabling legislation proved to be far-reaching as each act gave impetus to emerging policies of forestland conservation, management, renewal, and wise use. The concept of multiple-use became a working tool throughout the National Forest System and carried over into the private sector as well, setting the stage for forestland renewal.

Cooperative forestry programs provided a major boost to high-tech industrial forestry as we know it today. They also helped elevate our profession of forestry to worldwide standards of distinction. And, of course, they did a great deal to elicit much needed support for the emerging conservation ethic, which in 1875 was only the dream of a very small group.

PART IV

FINDING COMMON GROUND

9

HAVE YOU CARED FOR
A TREE TODAY?

THE DIFFICULT PART of our quest for common ground will be exposing the misleading perceptions that literally surround us. For example, in New England a news article in 1991 contained pictures of college students carrying signs protesting, of all things, tree farms.

In another example, the *Providence Journal Bulletin* of 2 April 1994 reported that an environmental curriculum distributed by Proctor & Gamble to seventy thousand teachers said, "Clearcutting. . . opens the forest floor to sunshine, thus stimulating growth and providing food for animals." This statement was criticized by "environmental experts" who said that "clearcutting. . . denudes swathes of forest. . . [causing] erosion, loss of wildlife habitat and the replacement of vegetation with single-species tree farms."

The logic of these "experts" is wearing thin. Harvesting, the first step in renewal, does not denude forestland, nor does healthy forestland cause erosion. Neglect does. The Missouri River, the "Big Muddy," was beset by erosion long before the colonists arrived. And of course, there is no such thing as "single-species tree farms" if you look at creditable scientific evidence that reveals hundreds of diverse species of wildlife and plant life.

Swarthmore College students in Pennsylvania successfully opposed a forest sale on the Monongahela National Forest in West Virginia. In an April 1992 news article, one student appraising forestland practices in the public sector said, "The Forest Service is creating economic markets that are destroying the natural forests." Students objected to "the construction of new roadways," and to "destroying the 'biological integrity' of the forest [which] would sacrifice the environment to business interests." They were further quoted as saying, "The taxpayers' money pays for the assessments and the drawing up of the plans, and the private companies reap the profits."

Here is yet another instance of well-meaning students making a sincere effort to do good. Unfortunately, their efforts are based on feelings rather than facts—they have little or no understanding about the resilient and dynamic nature of our renewable forestland ecosystems.

It is also unfortunate that environmental case histories are inundated by negative commentary in the media. In a March 1992 Op Ed article, "Forest for Sale: It's a Steal," the writer informs us:

> When clearcut forests are replanted, and if the replanting takes, the result is a tree farm, not a forest. Biologically impoverished monocultures replace diverse ecosystems; habitat for fish and wildlife is severely damaged or destroyed. Of many species, few remain. . . . At stake is a biomass of a density equal to that of a tropical rain forest, and of similar importance to the planet's life-support system. Mass clearcutting destroys lands and watersheds and causes desertification for many miles downwind. Fish and wildlife disappear.[1]

As with the young people from Swarthmore, this writer means well, but like the students, he is a victim of misinformation.

The misinformation in the media and in literature is endless. For example, in the June 1991 *Earthwatch Magazine*, a letter to the editor had this to say:

> I am outraged that *Earthwatch* printed an ad by the American Forest Council. The ad is full of slippery half-truths designed to promote the malicious agenda of the timber industry under the guise of responsible forestry.

The council proclaims that billions of seedlings are planted every year. This is true. But these single-species plantings follow the clearcutting of irreplaceable forest ecosystems that have evolved over thousands of years. It's ironic that their ad ran in the same issue that discussed the plight of the spotted owl, one of many species threatened by the "forest management" practices endorsed by the council.

By advertising in *Earthwatch*, the timber industry is using your reputation to lend credence to its reckless program.[2]

The "reckless program" in question is, of course, the scientific renewal of our forestland ecosystems that has proved to be so successful.

CARING IN THE REAL WORLD

We are also literally surrounded by plenty of opportunities to become advocates of exemplary forestland conservation. There are indeed signs that misleading perceptions are wearing thin; both students and educators are coming to the fore. In early October 1991, some 2,200 students attended a weekend National Student Environmental Conference on the campus of the University of Colorado in Boulder.

Here, according to the *New York Times* (7 October 1991), they "called for a new agenda for the American environmental movement, one that joins issues of race, class, and injustice with the movement's traditional goals of preservation and conservation." According to the news article, "At issue is what student leaders here said was the narrow objectives of government environmental agencies and the major national environmental groups."

The conference was the first of two national meetings "to focus on what organizers call the elitism of the American environmental movement." One coordinator for the student coalition, Randolph Viscio, offered: "Poor housing is an environmental issue." Another, Miya

Yoshitani, said: "We wanted a movement that can make a difference, and the only way we can achieve success is to make it inclusive and broad-based. . . .Is it right that some students are more willing to drive two hours to save a forest than to drive five minutes to help people in a poor community?"

These young people called on the magic of the American dream. Nothing is more positive than the philosophy and courage of these students. By focusing their energy on complex environmental issues for the benefit of *all* citizens rather than a privileged few, they bring into play the American generosity and courage that are unfailing in times of crisis and risk. And you can be sure our renewable forestland ecosystems are at risk.

It's possible that in part the inspiration of these college students stems from the American Forest Foundation award-winning Project Learning Tree (PLT). Implemented through a volunteer network, PLT (designed for grades kindergarten through 12) is active in all fifty states, four U.S. territories, and other countries. Over the past twenty years, more than 400,000 educators have been trained to use its thematic, interdisciplinary curriculum reaching an estimated 20 million students.

The newest version of PLT released in 1994 provides teachers, students, and parents with opportunities to investigate key issues and make informed, responsible decisions about the environment.

State coordinators and steering committees selected ten outstanding teachers who exemplify the commitment and dedication to further environmental education in their classrooms and communities through the use of the PLT curriculum. Invited to a reception held in Washington, D.C., in their honor in May 1994 were:

Katherine Snyder, Arkansas	Joanne Alex, Maine
Barbara Leonard, Colorado	Ruth Carlstrom, Montana
Myra Jeffres, FLorida	Dr. Christine Moseley, Oklahoma
Chris Johnson, Georgia	Wesley Roberts, Tennessee
Willa Dietz, Louisiana	Pat Lisoskie, Washington

Caring: Deep Creek Lake State Park

There is no greater satisfaction than playing an active and positive role in renewing our forestland ecosystems. There are many low-profile heroes who are doing just that. In the spring of 1991, twenty of these heroes (loggers and foresters) came to the rescue of Maryland's Deep Creek Lake State Park.

This park, near McHenry—located about 35 miles west of Cumberland, some 135 miles west of Baltimore—receives little publicity. Funds for effective pest control have been inadequate and the oak forest stands in the park and throughout the region have been devastated by continuous infestations of the gypsy moth. But whereas other Maryland state parks were slated for closing because of budget shortfalls, Deep Creek Lake was not.

Park officials were greatly concerned, however, that the towering oak snags were a deadly hazard in the day-use area. In order to be able to open on Memorial Day weekend, it was estimated that some 250 of these snags would have to be removed at a cost of over $52,000. To the rescue came the Maryland Forests Association.

As a free public service, twenty members of this private sector forestry organization removed all of the snags with little residual damage and minimal park disturbance. The volunteers contributed some two hundred man-hours, and the salvaged oak logs were manufactured into lumber at the state's sawmill for use at other parks.

In addition, over three hundred cords of firewood were made available for sale to generate funds to purchase replacement trees. The park opened on schedule and seasonal gate receipts in excess of $100,000 were assured.

Donna and I observed the handiwork of these volunteers in the fall of 1992.

Anyone visiting any part of our Eastern forestlands when fall colors dazzle the senses can attest to the many-faceted beauty of the scene, the sight of a sweep of color. But this park, in any season, is a splendor of color and light.

The facilities are well planned and cared for. Whether your interests are overnight camping, boating, picnicking, bird watching, or just a change from the humdrum of daily living, this park fulfills the wish.

A glance at the playground area in photo 38 will give you an idea of why parents and children find the park attractive. Note that what remains of the dangerous snags are groundlevel stumps that you can see in the foreground of the photo.

Logic, a subject which antiforestry advocates would like to avoid, demanded the removal of these deadly hazards for the safety of children, parents, and other visitors in this park. Why, then, shouldn't it also be true for the safety of Forest Products Industry employees, hunters, and recreationists?

And why are state and federal legislators passing laws that mandate leaving snags in both public and private sector forestlands? If logic has any meaning at all, how do these laws relate to forestland conservation, management, renewal, and wise use in the real world?

Caring: Calaveras Big Trees State Park

While such positive case histories as the Deep Creek Lake State Park receive little publicity, they are widespread. A problem similar to the one it had overcome faced state park officials in California.

The Calaveras Big Trees State Park is located in the Sierra foothills, halfway between Lake Tahoe to the north and Yosemite National Park to the south, about 160 miles east of San Francisco. In the fall of 1990, at the behest of park officials, some two hundred dead and dying trees, some as tall as twenty story buildings, were removed from within the main campground area by a local "one-man" logging company. By the summer of 1991, North Grove visitors to the park were greeted with a little more sunlight, no hazardous snags, and beautifully landscaped campsites.

In the Sierra pine country, as in Eastern hardwood forestlands, beauty and safety are often subjective. While it may be appropriate occasionally to leave dangerous snags, such as in certain research areas and unique Wilderness set-asides, it makes little sense to leave them in areas frequented by the general public. Surely, it's time to encourage the sort of forestland conservation that includes common sense stewardship of our public as well as private forestlands.

Citizen conservationists can participate in state park projects such as these in Maryland and California, at least to some extent. When speaking of ordinary citizens, the heavy work—felling, bucking, skidding, and hauling—must be left to professionals; but the cosmetics, clean-up chores, and tree planting require just as many or more hours of labor. The renewal of the forestland is the most important work of all. And for this, legions of volunteers are required.

Muir Woods National Monument, just a few miles north of San Francisco, is a wonderful example of a successful volunteer program. This park has the appearance of a Garden of Eden, yet it is one of the most heavily visited of all the redwood parks. Volunteers deserve much of the credit.

This also holds true of the "desert oasis" of Sabino Canyon just outside Tucson (USDA, Santa Catalina Ranger District, Coronado National Forest). Inside the visitors center is a plaque recognizing thousands of hours of service by a cadre of volunteers.

But back to the Calaveras Big Trees State Park. Since it was the last stop on the photo expedition mentioned in Chapter 3, some of the history of this unusual park seems appropriate here.

Its name is derived from the giant sequoias found in the roadless South Grove, which runs northeast 3 miles along Big Trees Creek varying in elevation from 4,500' to 5,500'. When the South Grove was acquired in 1929 by the Pickering Lumber Company, the Calaveras Grove Association became active and was instrumental in establishing the North Grove state park in June 1931.

Following World War II, Pickering revealed plans for harvesting the sugar pine stands that are now within the South Grove. It was never Pickering's intention to disturb the giant sequoias; nevertheless, this news alarmed citizen conservationists, who successfully raised $2.8 million to purchase the South Grove from Pickering in 1954. The South Grove was designated a Natural Preserve—similar to a federal Wilderness designation—by the State Park Commission in 1984.

A paved road runs through the North Grove for 10 miles to the South Grove trailhead. This trail provides access to the lower half of the South Grove, including the "Palace Hotel" sequoia and the "Agassiz," which is the largest sequoia in the grove and one of the ten largest sequoias in existence.

After World War II, Pickering successfully bid on a 379-acre stand of sugar pine located on National Forest land along the ridge between the Big Trees drainage in the South Grove and Beaver Creek drainage in the North Grove. Interestingly, these acres were also a high priority for the citizen conservationists. Through their efforts, this small patch of forestland became the Calaveras Big Trees National Forest on 29 January 1953, an unofficial addition to the South Grove.

In order to do this, the Forest Service and Pickering had to make a swap. For the swap, Pickering had to cruise the 379 acres in the summer of 1951. The high value of the sugar pine stand necessitated a 100-percent cruise; every tree was to be tallied. For the field work, Pickering retained forestland consultant Henry Thomas from Portland, Oregon. Thomas, in turn, retained Porteous & Company in Seattle, my employer at the time. I had seen a multitude of beautiful trees by that time but nothing had prepared me for this assignment. These sugar pines, remarkable in form and size, were pieces of art.

Pickering was running a railroad show and our base was their Skull Creek logging camp on a rail siding about a mile due south of the South Grove. To reach the sugar pine stand along the ridge line we had about a 2 mile hike back and forth through the South Grove each day. This, my first view of a sequoia grove, left me breathless. And the sugar pine stand, without a doubt, was the finest anywhere. Our twelve-log volume tables would tally up to 192' (16' logs in pine country). This height was exceeded tree after tree, day after day—the most beautiful trees I have ever seen.

John Muir describes the sugar pine in his book, *The Mountains of California*, originally published in 1894. "This is the noblest pine yet discovered, surpassing all others not merely in size but also in kingly beauty and majesty. It towers sublimely from every ridge and canyon of the range, at an elevation of from three to seven thousand feet above the sea, attaining most perfect development at a height of about 5,000 feet."[3]

In 1991, I was as unprepared for what I saw on my return to take photos as I had been at the time of my initial visit. The contrast was appalling, the deterioration of these majestic trees heartrending!

Preserving a portion of our redwoods, our sequoias, and the unique stands of other species is a good idea, but the antiforestry mania of "saving" tens of millions of acres of commercial forestland and placing them in designated reserves makes little sense. Trees cannot be saved in this manner, as is sadly evident when viewing these magnificent sugar pines and the overmature Douglas-fir stands in the Pacific Northwest.

These photos, taken when California was suffering a severe drought, show how badly distressed Sierra forestlands have become. The drought only hastens the inevitable. Even though the sugar pines we cruised in 1951 were overmature at the time (they were actually overmature at the turn of the century), mortality was not conspicuous.

Today, mortality among these beautiful trees tears at your heart. There is no way to prepare for the shock of seeing so many snags instead of lovely, live trees. On the higher elevations, above 5,000', mortality is running upwards of 25 percent.

- (Photo 39): This view is just inside the South Grove at the lower end of the ridge between the Beaver Creek drainage to the north and the Big Trees Creek drainage to the south. This series of photos follows a fire road that runs up the back of the ridge and around the northeast end of the South Grove. The South Grove is off to the right of the fire road and the Calaveras Big Trees National Forest is off to the left.

  Before a sugar pine behind the photographer caught her attention, our granddaughter, Sarah, was attracted to a tree more her size, a shade tolerant white fir seedling. After this photo she was ready to look into a playhouse in the forest.

- (Photo 40): Sarah is just to the left of the doorway to her sugar pine playhouse.

- (Photo 41): Sarah, mother Rita Lowell, and grandmother

(Donna) in the playhouse doorway. To give you an idea of the size, four people could sleep comfortably inside. Let us hope Sarah's sugar pine playhouse will still be there when she returns with her children.

- (Photo 42): This view is also on the border of the South Grove at the lower end of the ridge. And this sugar pine is an example of what I mean by badly distressed. Though not easy to see in the photo, the mortality is obvious when you are on the ground looking in all directions. There are two tremendous snags in the background of the photo.

- (Photo 43): This gives a better idea of scale. I am over 6 feet tall and this sugar pine was well over 200 feet tall.

- (Photo 44): We have moved up the fire road along the ridge and are now over 5,000' in elevation. This lone, live sugar pine surrounded by snags is right on the border of the Calaveras Big Trees National Forest. It is a tiny portion of the surrounding desolation of what, at the turn of the century, was a pine stand of "kingly beauty and majesty"—now a vivid scene of dying, dead, and down trees that can bring tears, and do, to the eyes.

- (Photo 45): This view, near the previous photo—a snag surrounded by live trees—is actually more representative of the mortality.

CARING: CAN YOU MAKE A DIFFERENCE?

If we are to continue to renew and care for our forestland ecosystems in the twenty-first century, far more citizen conservationists will be required than ever before and they will need to be able to distinguish between perceptions and reality. And the support of professional foresters and other ecosystem scientists will of course be required.

Only informed citizens on the march can make a difference in government, and therefore in the care of trees and forestland. As of 1993, 129,000 people were busily writing federal regulations. The fruits of their labor tower above us—66,000 pages of federal regulations. Any wonder that gridlock dominates our attempts to care for our forestland ecosystems?

Obviously, we should place a high priority on education and research. Professional forestry is a noble calling, the cornerstone to understanding our forestland ecosystems. Therefore, we should everywhere encourage a young person to take up forestry and environmental studies.

For information on the profession of forestry you may contact the Director of Science and Education, Society of American Foresters, 5400 Grosvenor Lane, Bethesda, MD 20814-2198, phone, 301-897-8720. In Table 12 in the appendix, all members of the National Association of Professional Forestry Schools and Colleges are listed and those accredited by the Society of American Foresters are noted. Table 13 includes technician schools accredited by the society.

In Table 14 in the appendix, the so-called "Group of 10" can be found. A number of these conservation organizations have regional offices as well as regional representatives which are also listed. The Group of 10 are the largest, most influential "environmental activist" organizations. Much that they do has great value and many well-meaning and well-intentioned—and even well-informed—people participate in these organizations.

Unfortunately, perhaps tragically, this group's leaders with few exceptions have taken a fork in the road away from the long-established conservation ethic for trees and forestlands. Since antiforestry rhetoric is the basis of their fund raising, this fork in the road is taking these organizations farther and farther away from the desirable goal of reaching common ground.

The antiforestry rhetoric of this group notwithstanding, much good information can be found in their publications. For readers desiring more detail on who's who in the fast-growing conservation industry, the National Wildlife Federation's annual *Conservation Directory* is the best single source.

In numbers, the federation is the leading conservation organization in the country. A good example of why they attract citizen conservationists is the Federation Pledge, quoted on the back of the directory:

I pledge myself, as a responsible human, to assume my share of the stewardship of our natural resources.

I will use my share with gratitude, without greed or waste.

I will respect the rights of others and abide by the law.

I will support the sound management of the resources we use, the restoration of the resources we have despoiled, and the safe-keeping of significant resources for posterity.

I will never forget that life and beauty, wealth and progress, depend on how wisely we use these gifts. . . . the soil, the water, the air, the minerals, the plant life and the wildlife.

THIS IS MY PLEDGE!

The best way to seek the truth regarding renewing and caring for our forestland ecosystems is to review the abundance of information available not only from these organizations but from other sources free from the antiforestry disease.

For fifty years, the supply miracle and the remarkable success of renewing our forestlands have gone unheralded by the general public. Information about this success story is available from the American Forest Foundation, 1111 19th Street NW, Suite 780, Washington, DC 20036, phone, 202-463-2462.

This foundation sponsors the American Tree Farm System as well as Product Learning Tree (PLT). Over seventy thousand members of this system care for some 93 million acres of forestland. These are the people who will carry into the twenty-first century the conservation ethic for trees and forestlands and they should be accorded the recognition and the gratitude they deserve for the contribution they are making to our well-being.

As a step in that direction each year the foundation selects the tree farmer of the year. Invited to Washington, D.C., in May 1994 to a reception in their honor were Wayne and Colleen Krieger from Gold Beach, Oregon. On their 310 acre Skyview Ranch overlooking the Pacific Ocean the Krieger family nurtures white fir, Douglas fir alder, myrtle, hemlock, cedar, oak, and redwood. They are also developing an educational forestry center to encourage environmental stewardship.

The membership of the American Tree Farm System includes both industrial and nonindustrial forestland owners. In 1983, the National Woodland Owners Association was founded to serve private, nonindustrial forestland owners. This organization, 374 Maple Avenue East, Suite 200, Vienna, Virginia 22180, phone, 703-255-2700, provides a networking opportunity for shared vested interests to over thirty thousand members in all fifty states. There are twenty-six state affiliates, which are listed in Table 15 in the appendix.

Information about cooperative forestry and multiple-use programs of the USDA Forest Service is available from the Public Affairs Office, 14th & Independence SW (Box 96090), Washington, DC 20090-6090, phone, 202-205-0975 or from any of the field offices listed in Table 16 in the appendix. Information is also available at each of the National Forest Supervisors' offices and their respective ranger stations which are listed in the local telephone directory.

Also listed in the appendix are professional forestry organizations with on-going grass roots programs for forestland conservation and management. These include the National Association of State Foresters for the public sector (Table 17) and the National Council of Forestry Association Executives for the private sector (Table 18).

The National Arbor Day Foundation, 100 Arbor Avenue, Nebraska City, NE 68410, phone, 402-474-5655, with its million plus members is the leading organization encouraging people to plant and care for trees.

For overall support of forestland conservation, management, renewal, and wise use, I recommend the American Forestry Association, 1516 P Street, NW, Washington, DC 20005, phone, 202-667-3300. In recent years, its emphasis has been on urban forestry. Under its present leadership, industrial and nonindustrial forestry programs have not been a priority. Leaders in the Forest Products Industry and the leaders of this fine conservation organization must take it upon themselves to seek common ground and pull together more effectively as we enter the coming century.

Also worthy of mention is one of the more recent conservation organizations, the National Wilderness Institute (NWI). In contrast to the Group of 10, its positive agenda of advancing a correct understanding of the inherent dynamic and resilient characteristics of our renewable resources places it in the vanguard of helping citizens to deal with forestland conservation and management.

The philosophy of NWI is based on sound science and fact, not on hype, hysteria, and worst-case scenarios. NWI has conducted extensive research on endangered species, wetlands, and the National Biological Survey and the staff speaks regularly to university and community groups and professional associations. The motto of this group, "Voice of Reason on the Environment," points us in the right direction. For more information, call 703-836-7404 or write 25766 Georgetown Station, Washington, DC 20007.

THE IMPACT OF CARING

In 1881, when Oberforster Richard Baron von Steuben journeyed to Cincinnati and convinced newspaperman William L. DeBeck that forestland conservation was a subject "urgently needing public attention," the result was the first American Forest Congress (described in Chapter 6). This momentous occasion lasted five days and had enormous impact. It focused the attention of the public on forestland conservation as has no other event in our history. Fifty thousand spectators attended the various events.

It is hard to imagine Oberforester von Steuben's reaction were he to view our distressed and neglected forestlands today. Gridlock has separated foresters and forestlands in the public sector, and Congress has paid little attention to the historic conservation ethic for trees and forestlands and to the charter of the Forest Service that seeks to supply the needs of all citizens.

What can be done about this dilemma? If correct, Drucker's prediction—that there will be too few challenging positions for volunteers—surely offers a clue. It is time for Congress, the media, environmental community leaders, Forest Products Industry leaders, and true conservationists to do some serious soul-searching. But above all, it is the lay people who can break the gridlock.

Unfortunately, lay people do not seem to understand the importance of their role in reversing the simplistic notions that have led to gridlock and neglect of our renewable forestland ecosystems. It is my hope that this book will give both professional resource managers and lay people helpful insights into the contentious issues that beset us and awaken their enthusiasm for a renewed conservation ethic for trees and forestlands.

There is no question that we need renewable forestland ecosystems, and for many reasons. We need them now and we will need them in the future and forever. We—people of good will everywhere—should and will come together and care for our trees and forestlands, and when we do we will have an enormous impact. Let us begin today.

APPENDIX

Table 8
LAND SET ASIDES BY STATE (000 ACRES)

State	National Forest Service	Bureau of Land Management	Fish and Wildlife Service	National Park Service	Total
Alabama	42	—	53	7	102
Alaska	9,072	7,362	76,385	55,181	148,000
Arizona	2,145	2,160	1,707	2,830	8,842
Arkansas	262	—	258	105	625
California	5,003	8,270	365	5,010	18,648
Colorado	2,650	1,041	57	603	4,351
Connecticut	—	—	—	8	8
Delaware	—	—	25	—	25
Florida	154	—	534	2,629	3,317
Georgia	172	—	477	64	713
Hawaii	—	—	271	259	530
Idaho	5,034	2,563	80	106	7,783
Illinois	26	—	98	—	124
Indiana	13	—	8	14	35
Iowa	—	—	73	2	75
Kansas	—	—	56	1	57
Kentucky	17	—	2	94	113
Louisiana	91	—	442	21	554
Maine	12	—	43	85	140
Maryland	—	—	37	69	106
Massachusetts	—	—	12	56	68
Michigan	122	—	115	716	953
Minnesota	1,087	—	206	247	1,540
Mississippi	11	—	181	116	308
Missouri	78	—	46	83	207
Montana	3,494	498	1,126	1,273	6,391
Nebraska	15	—	147	6	168
Nevada	788	5,152	2,312	778	9,030
New Hampshire	103	—	3	10	116
New Jersey	—	—	56	53	109
New Mexico	1,408	1,251	384	389	3,432
New York	—	—	25	95	120
North Carolina	210	—	414	386	1,010
North Dakota	—	—	290	73	363

State	National Forest Service	Bureau of Land Management	Fish and Wildlife Service	National Park Service	Total
Ohio	—	—	8	33	41
Oklahoma	42	—	153	10	205
Oregon	2,649	3,240	567	198	6,654
Pennsylvania	32	—	10	109	151
Rhode Island	—	—	1	—	1
South Carolina	74	—	176	28	278
South Dakota	44	—	48	274	366
Tennessee	77	—	110	369	556
Texas	36	—	417	1,221	1,674
Utah	870	3,411	95	2,096	6,472
Vermont	96	—	6	8	110
Virginia	275	—	119	336	730
Washington	2,745	5	173	1,945	4,868
West Virginia	181	—		80	261
Wisconsin	107	—	161	134	402
Wyoming	3,224	1,076	64	2,396	6,760
TOTAL	42,461	36,029	88,396	80,606	247,492

Sources: Forest Service, USDA, FS 383, 1993
Land Areas of the National Forest System
As of 30 September 1992

USDI, Bureau of Land Management
Public Land Statistics 1990, Volume 175
August 1991

USDI, U.S. Fish and Wildlife Service
Annual Report of Lands Under Control of the U.S. Fish and Wildlife Service
As of 30 September 1992

USDI, National Park Service
As of 30 September 1992

Table 9
FEDERAL LAND BY STATE (000 ACRES)

State	Total State	Total Federal	Percent Federal
Alabama	32,678	550	1.7
Alaska	365,482	247,802	67.8
Arizona	72,688	31,491	43.3
Arkansas	33,599	3,421	10.2
California	100,207	61,043	60.9
Colorado	66,486	22,648	34.1
Connecticut	3,135	14	.4
Delaware	1,266	30	2.4
Florida	34,721	3,355	9.7
Georgia	37,295	2,292	6.1
Hawaii	4,106	677	16.5
Idaho	52,933	33,122	62.6
Illinois	35,795	494	1.4
Indiana	23,158	470	2.0
Iowa	35,860	159	.4
Kansas	52,511	690	1.3
Kentucky	25,512	1,391	5.4
Louisiana	28,868	6,538	22.6
Maine	19,848	153	.8
Maryland	6,319	197	3.1
Massachusetts	5,035	83	1.6
Michigan	36,492	3,565	9.8
Minnesota	51,206	2,387	4.7
Mississippi	30,223	1,670	5.5
Missouri	44,248	2,030	4.6
Montana	93,271	25,862	27.7
Nebraska	49,032	719	1.5
Nevada	70,264	57,803	82.3
New Hampshire	5,769	754	13.1
New Jersey	4,813	135	2.8
New Mexico	77,766	25,747	33.1
New York	30,681	223	.7
North Carolina	31,403	1,141	3.6
North Dakota	44,452	1,965	4.4
Ohio	26,222	322	1.2

State	Total State	Total Federal	Percent Federal
Oklahoma	44,088	874	2.0
Oregon	61,599	29,669	48.2
Pennsylvania	28,804	641	2.2
Rhode Island	677	5	.7
South Carolina	19,374	434	2.2
South Dakota	48,882	2,744	5.6
Tennessee	26,728	1,322	4.9
Texas	68,218	2,845	1.7
Utah	52,697	33,611	63.8
Vermont	5,937	355	6.0
Virginia	25,496	1,918	7.5
Washington	42,694	12,373	29.0
West Virginia	15,411	2,099	13.6
Wisconsin	35,011	1,906	5.4
Wyoming	62,343	30,407	48.8
TOTAL	2,271,342	662,157	Avg 16.0

Source: USDI, Bureau of Land Management
Public Land Statistics 1990, Volume 175
August 1991

Table 10
FOREST PRODUCTS INDUSTRY SELECTED LIST OF
POTENTIAL INFORMATION CENTER INVESTORS
1991 Sales and Net After Tax Profit ($000,000)

GROUP I $2 BILLION PLUS		Sales	Profit
1. IP	Purchase, NY	$12,700	$184
2. Georgia-Pacific	Atlanta, GA	11,524	(142)
3. Weyerhaeuser	Tacoma, WA	8,702	(162)
4. Kimberly-Clark	Dallas, TX	6,777	508
5. Stone Container	Chicago, IL	5,384	(49)
6. Scott	Philadelphia, PA	4,976	(70)
7. Champion	Stamford, CT	4,786	40
8. Mead	Dayton, OH	4,579	7
9. James River	Richmond, VA	4,651	78
10. Boise Cascade	Boise, ID	3,950	(79)
11. Union Camp	Wayne, NJ	2,967	125
12. Temple-Inland	Diboll, TX	2,507	138
13. *Westvaco	New York, NY	2,301	136
14. Willamette	Portland, OR	2,004	46
GROUP I TOTAL		$77,808	$760

GROUP II $3/4 BILLION - $2 BILLION			
15. L-P	Portland, OR	$1,702	$56
16. Sonoco	Hartsville, SC	1,697	95
17. Federal Paper Board	Montvale, NJ	1,435	82
18. Bowater	Darien, CT	1,288	46
19. Potlatch	San Francisco, CA	1,237	56
20. Pentair	St. Paul, MN	1,169	41
21. Consolidated Papers	Wisconsin Rapids, WI	872	91
22. Chesapeake	West Point, VA	840	15
GROUP II TOTAL		$10,240	$482

GROUP III $100 MILLION - $3/4 BILLION			
23. *Longview Fibre	Longview, WA	$644	17
24. Glatfelter	Spring Grove, PA	568	74
25. Pope & Talbot	Portland, OR	502	(12)
26. *Wausau Paper	Wausau, WI	350	30
27. *Fiberboard	Concord, CA	245	1
28. *WTD	Portland, OR	244	(85)
29. Mosinee	Mosinee, WI	197	1

30. Rayonier Timberland	New York, NY	101	73
GROUP III TOTAL		2,851	99
GROUP I, II & III TOTAL		$90,899	$1,341

* Not on calendar fiscal year

Source: WSJ, HMA

Table 11

FOREST PRODUCTS INDUSTRY SELECTED LIST OF
POTENTIAL INFORMATION CENTER INVESTORS
FIVE YEAR SALES AND PROFIT, 1987 TO 1991
Ranked by 1991 sales ($000,000)

Company	Year	Sales	Profit	Year	Sales	Profit
GROUP I $2 BILLION PLUS						
1. IP	1987	7,800	407	1990	13,000	569
	1988	9,500	754	1991	12,700	184
	1989	11,400	864			
2. G-P	1987	8,603	458	1990	12,665	365
	1988	9,509	467	1991	11,524	(142)
	1989	10,171	661			
3. Weyerhaeuser	1987	8,823	447	1990	9,024	394
	1988	9,328	566	1991	8,702	(162)
	1989	10,106	341			
4. Kimberly Clark	1987	4,900	325	1990	6,407	432
	1988	5,393	379	1991	6,777	508
	1989	5,734	424			
5. Stone Container	1987	3,233	161	1990	5,756	95
	1988	3,742	342	1991	5,384	(49)
	1989	5,330	286			
6. Scott	1987	4,122	234	1990	5,356	148
	1988	4,726	401	1991	4,976	(70)
	1989	5,066	375			
7. Champion	1987	4,615	382	1990	5,090	223
	1988	5,128	456	1991	4,786	40
	1989	5,163	432			
8. Mead	1987	4,209	217	1990	4,772	38
	1988	4,464	353	1991	4,579	7
	1989	4,608	216			
9. James River	1987	4,479	170	1990	5,416	78
	1988	5,098	209	1991	4,562	78
	1989	5,872	255			
10. Boise Cascade	1987	3,821	183	1990	4,186	75
	1988	4,095	289	1991	3,950	(79)
	1989	4,338	268			

Company	Year	Sales	Profit	Year	Sales	Profit
11. Union Camp	1987	2,362	207	1990	2,840	230
	1988	2,661	295	1991	2,947	125
	1989	2,761	299			
12. Temple Inland	1987	1,603	142	1990	2,401	232
	1988	2,017	199	1991	2,507	138
	1989	2,124	207			
13. Westvaco	1987	1,904	146	1990	2,411	188
	1988	2,134	200	1991	2,301	137
	1989	2,284	223			
14. Willamette	1987	1,432	121	1990	1,905	130
	1988	1,716	161	1991	2,004	46
	1989	1,892	191			

GROUP II $3/4 BILLION - $2 BILLION

Company	Year	Sales	Profit	Year	Sales	Profit
15. L-P	1987	1,654	125	1990	1,793	91
	1988	1,799	300	1991	1,702	56
	1989	2,009	193			
16. Sonoco	1987	1,312	61	1990	1,669	50
	1988	1,600	96	1991	1,697	95
	1989	1,666	104			
17. Fed Paper Bd	1987	1,026	66	1990	1,374	118
	1988	1,117	143	1991	1,435	82
	1989	1,310	205			
18. Bowater	1987	1,231	81	1990	1,380	78
	1988	1,410	164	1991	1,288	46
	1989	1,450	145			
19. Potlatch	1987	992	88	1990	1,253	99
	1988	1,084	112	1991	1,237	56
	1989	1,228	137			
20. Pentair	1987	789	22	1990	1,176	33
	1988	823	40	1991	1,169	41
	1989	1,164	36			
21. Consol Papers	1987	743	96	1990	949	142
	1988	897	150	1991	872	91
	1989	952	168			
22. Chesapeake	1987	676	30	1990	841	17

Company	Year	Sales	Profit	Year	Sales	Profit
	1988	711	51	1991	840	15
	1989	814	48			
23. Longview Fibre	1987	569	65	1990	685	61
	1988	657	96	1991	644	17
	1989	698	66			
24. Glatfelter	1987	424	55	1990	625	88
	1988	569	82	1991	568	76
	1988	599	93			
25. Pope & Talbot	1987	427	28	1990	562	20
	1988	515	32	1991	502	(12)
	1989	619	44			
26. Wausau Paper	1987	251	12	1990	340	16
	1888	284	16	1991	350	30
	1989	317	21			
27. Fiberboard	1987	154	8	1990	245	1
	1988	175	12	1991	234	(44)
	1989	218	5			
23. WTD	1987	294	8	1990	244	(85)
	1988	360	9	1991	214	3
	1989	460	1			
29. Mosinee	1987	216	6	1990	210	12
	1988	232	9	1991	197	1
	1989	233	7			
30. Rayonier Timberland	1987	69	46	1990	106	79
	1988	91	76	1991	101	73
	1989	95	72			

ANNUAL RECAP

	Year	Sales	Profit	Percent
	1987	72,733	4,397	6%
	1988	81,835	6,459	8%
	1989	90,681	6,387	7%
	1990	94,681	4,107	4%
	1991	91,347	1,300	1%
Total		$431,277	$22,650	5%

Source: WSJ, HMA

Table 12
NATIONAL ASSOCIATION OF PROFESSIONAL
FORESTRY SCHOOLS AND COLLEGES

Emmett F. Thompson
Dean, School of Forestry
* Auburn University
108 M. White Smith Hall
Auburn, AL 36849-5418
205-844-1007

James V. Drew
Dean, School of Agriculture & Land
Resource Management
Director, Agricultural
Experiment Station
University of Alaska, Fairbanks
172 AHRB
Fairbanks, AK 99775-0100
907-474-7083

Lawrence D. Garrett
Dean, School of Forestry
* Northern Arizona University
Box 4098
Flagstaff, AZ 86011
602-523-3031

C. P. P. Reid, Director
School of Renewable Resources
Bioscience E. Rm. 325
University of Arizona
Tucson, AZ 85721
602-621-7257

John Gunter, Dean
School of Forest Resources
* University of Arkansas
Box 3468
Monticello, AR 71655
501-460-1052

Norman H. Pillsbury
Head, Natural Resources Management
Department
** California Polytechnic State University
San Louis Obispo, CA 93407
805-756-2702

James P. Smith, Dean
College of Natural Resources & Sciences
* Humboldt State University

Arcata, CA 95521
707-826-3256

James Bartolome
Dept. of Env. Science, Policy & Mgmt.
* University of California, Berkeley
145 Mulford Hall
Berkeley, CA 94720
510-642-0376

A. A. Dyer, Dean
College of Natural Resources
101 Natural Resources Bldg.
* Colorado State University
Ft. Collins, CO 80523
303-491-6675

John F. Anderson, Director
Connecticut Ag. Experiment Station
Box 1106
New Haven, CT 06504
203-789-7214

David B. Schroeder, Head
Dept. of Natural Resources,
  Mgmt and Engineering
College of Agriculture & Natural
  Resources
University of Connecticut
1376 Storrs Road
Storrs, CT 06269-4087
203-486-2840

Jared L. Cohon, Dean
School of Forestry & Environmental
  Studies
* Yale University
205 Prospect Street
New Haven, CT 06511
203-432-5109

Dr. John C. Nye, Director
Agricultural Experiment Station
University of Delaware
133 Townsend Hall
Newark, DE 19717-1303
302-831-2501

Loukas Arvanitis, Interim Director
School of Forest Resources & Conservation
* University of Florida
118 Newins-Ziegler Hall
Box 110410
Gainesville, FL 32611-0410
904-392-1791

Arnett C. Mace, Jr., Dean
School of Forest Resources
* University of Georgia
Forest Resources Building, Room 231
Athens, GA 30602-2152
706-542-2686

Noel P. Kefford, Dean
College of Tropical Ag & Human Resources
University of Hawaii
3050 Maile Way, Room 202
Honolulu, HI 96822
808-956-8234

John C. Hendee, Dean
College of Forestry, Wildlife &
  Range Sciences
* University of Idaho
Moscow, ID 83843
208-885-6442

Dwight McCurdy, Chair
Department of Forestry
* Southern Illinois University
Carbondale, IL 62901
618-453-7474

Gary L. Rolfe, Head
Department of Forestry
* University of Illinois
West 503 Turner Hall
1102 South Goodwin
Urbana, IL 61801
217-333-2770

Dennis C. Le Master, Head
Dept of Forestry & Natural Resources
* Purdue University
1159 Forestry Bldg.
West Lafayette, IN 47907-1159
317-494-3590

Steven E. Jungst, Chairman
Department of Forestry
* Iowa State University

251 Bessey Hall
Ames, IA 50011
515-294-1166

Thomas D. Warner, Head
Dept of Horticulture, Forestry &
  Rec Resources
Kansas State University
266 Waters Hall
Manhattan, KS 66505
913-532-6170

Robert N. Muller, Chairman
Department of Forestry
* University of Kentucky
205 Thomas Poe Cooper Building
Lexington, KY 40546-0073
606-257-7596

Stanley B. Carpenter, Director
School of Forestry, Wildlife & Fisheries
* Louisiana State University
227 Forestry-Wildlife-Fisheries Building
Baton Rouge, LA 70803-6200
504-388-4227

G. H. Weaver, Director
School of Forestry
* Louisiana Tech University
Box 10138, Tech Station
Ruston, LA 71272
318-257-4985

G. Bruce Wiersma, Dean
College of Natural Resource, Forestry & Ag
* University of Maine
5782 Winslow Hall
Orono, ME 04469-5782
207-581-3202

T. J. Sexton, Acting Director
Agricultural Experiment Station
University of Maryland
3300 Metzerott Road
College Park, MD 20742
301-445-2665

Donald G. Arganbright, Head
Department of Forestry & Wildlife
  Management
* University of Massachusetts
Amherst, MA 01003
413-545-2665

Daniel E. Keathley, Chair
Department of Forestry
* Michigan State University
126 Natural Resource
East Lansing, MI 48824-1222
517-355-0093

Warren E. Frayer, Dean
School of Forestry & Wood Products
* Michigan Tech University
1400 Townsend Drive
Houghton, MI 49931
906-487-2915

Garry D. Brewer, Dean
School of Natural Resources
* University of Michigan
Ann Arbor, MI 48109-2195
313-764-2550

Alan Ek, Head
Dept of Forest Resources
* University of Minnesota
115 Green Hall
1530 Cleveland Avenue N.
St. Paul, MN 55108
612-624-3400

Al Sullivan, Dean
College of Natural Resources
University of Minnesota
235 Natural Resources Adm. Bldg.
2203 Upper Buford Circle
St. Paul MN 55108
612-624-1234

Warren S. Thompson
Dean, School of Forest Resources
* Mississippi State University
P.O. Drawer FR
Mississippi State, MS 39762-5726
601-325-2952

Albert R. Vogt, Director
School of Natural Resources
* University of Missouri
130 Agriculture Bldg.
Columbia, MO 65211
314-882-6446

Dean
School of Forestry
* University of Montana

Missoula, MT 59812
406-243-5522

Gary Hergenrader, Head
Dept of Forestry, Fisheries and Wildlife
University of Nebraska, East Campus
101 Plant Industry Building
Lincoln, NE 58583-0814
402-472-2944

G. F. Gifford, Chairman
Department of Range, Wildlife & Forestry
Mail Stop 186
University of Nevada
100 Valley Road
Reno, NV 89512-0013
702-784-4000

William W. Mautz, Chairman
Department of Forest Resources
* University of New Hampshire
215 James Hall
Durham, NH 03824
603-862-1020

Teuvo Airola, Assoc Prof.
Dept. of Natural Resources
Cook College, Rutgers University
Box 231
New Brunswick, NJ 08903
908-932-9631

Leroy Daugherty, Head
Horticulture Department
New Mexico State University
Box 3 Q
Las Cruces, NM 88003
505-646-3405

Daniel J. Decker, Chair
Department of Natural Resources
Cornell University
117 Fernow Hall
Ithaca, NY 14853-3001
607-255-2298

Bob G. Blackmon
Chairman, Faculty of Forestry
* SUNY College of Environmental
   Science and Forestry
Syracuse, NY 13210
315-470-6536

Norman L. Christensen, Dean
School of The Environment
* Duke University
216 Biological Sciences Building
Durham, NC 27706
919-684-2135

Larry W. Tombaugh, Dean
College of Forest Resources
* North Carolina State University
2028 Biltmore Hall, Box 8001
Raleigh, NC 27695-8001
919-515-2883

A. A. Boe, Chairman
Department of Horticulture & Forestry
North Dakota State University
Box 5658
Fargo, ND 58102
701-237-8162

Robert E. Roth, Acting Director
School of Natural Resources
Ohio State University
210 Kottman Hall
2021 Coffey Road
Columbus, OH 43210
614-292-2265

Edwin Miller, Head
Department of Forestry
* Oklahoma State University
011 Agricultural Hall
Stillwater, OK 74078
405-744-5437

George W. Brown, Dean
College of Forestry
* Oregon State University
Corvallis, OR 97331-5704
503-737-2221

Bart A. Thielges, Associate Dean
College of Forestry
Oregon State University
Peavy Hall, Room 154
Corvallis, OR 97331-5704
503-737-2222

Kim C. Steiner, Interim Director
School of Forest Resources
* Pennsylvania State University
113 Ferguson Building

University Park, PA 16802
814-863-7093

Jorge L. Rodriguez
College of Agricultural Sciences
University of Puerto Rico–
   Mayaguez Campus
College Station
Mayaguez, Puerto Rico 00708

Benton H. Box, Dean
College of Forest & Rec. Resources
130 Lehotsky Hall
* Clemson University
Clemson, SC 29634-1001
803-656-3215

W. Carter Johnson, Head
Dept of Horticulture, Forestry,
   Landscape & Parks
South Dakota State University
Box 2207-C
Brookings, SD 57007
605-688-5136

Dr. John Rennie, Acting Dept. Head
Department of Forestry, Wildlife & Fisheries
* University of Tennessee
Box 1071
Knoxville, TN 37901-1071
615-974-7126

R. Scott Beasley, Dean
School of Forestry
* Stephen F. Austin State University
Box 6109, SFA Station
Nocogdoches, TX 75962-6109
409-568-3304

Richard F. Fisher, Head
Department of Forest Science
* Texas A&M University
305 Horticulture/Forest Science Building
College Station, TX 77843-2135
409-845-5000

Dr. Terry L. Sharik, Head
Department of Forest Resources
* Utah State University
Logan, UT 84322-5215
801-750-2455

Lawrence K. Forcier, Dean
School of Natural Resources
* University of Vermont
330 Aiken Center
Burlington, VT 05401
802-656-4280

Gregory N. Brown, Dean
College of Forestry and Wildlife Resources
* Virginia Tech University
324 Cheatham Hall
Blacksburg, VA 24061-0324
703-231-5481

D. S. Padda, Vice President
Research and Land Grant Affairs
University of the Virgin Islands
Route 2, Box 10,000
Kingshill, St. Croix, VI 00850
809-778-0050

David B. Thorud, Dean
College of Forest Resources
* University of Washington
Seattle, WA 98195
206-685-1928

Edward Depuit, Chairman
Department of Natural Resource Sciences
* Washington State University
Pullman, WA 99164-6410
509-335-6166

Jack E. Coster, Director
Division of Forestry
* West Virginia University
Morgantown, WV 26506
304-293-2941

Donald R. Field, Director
School of Natural Resources
University of Wisconsin-Madison
146 Agriculture Hall
Madison, WI 53706
608-262-6968

Ronald L. Giese, Chairman
Department of Forestry
* University of Wisconsin
Russell Labs
Madison, WI 53706
608-262-1780

Alan Haney, Dean
College of Natural Resources
* University of Wisconsin
Stevens Point, WI 54481
715-346-2955

William Laycock, Head
Division of Range Management
University of Wyoming
Box 3354, Univ. Station
Laramie, WY 82071
307-766-2263

Source: National Association of Professionl Forestry Schools and Colleges
10 May 1994

* Society of American Foresters accredited schools

** SAF program candidate

Table 13
SOCIETY OF AMERICAN FORESTERS RECOGNIZED
FOREST TECHNICIAN SCHOOLS (TWO-YEAR ASSOCIATE DEGREE)

Lurleen B. Wallace State Junior College
Forest Technology Program
P.O. Drawer 1418
Andalusia, AL 36420

Kings River Community College
Forest and Park Technology Curriculum
995 North Reed Avenue
Reedley, CA 93654

Sierra College
Forest Technology Department
5000 Rocklin Road
Rocklin, CA 95677

Lake City Community College
Forest Technology Program
Route 7, Box 378
Lake City, FL 32055

Abraham Baldwin Agricultural College
Forestry/Wildlife Program
ABAC Station
Tifton, GA 31793

Allegany Community College
Forestry Program
Willow Brook Road
Cumberland, MD 21502

Michigan Technological University
Forest Technology Program
School of Technology
Houghton, MI 49931

Vermilion Community College
Natural Resource Program
1900 E. Camp Street
Ely, MN 55731

Flathead Valley Community College
Department of Forest Technology
One First Street E.
Kalispell, MT 59901

University of New Hampshire
Thompson School, Forest Technician
Program
Putnam Hall
Durham, NH 03824

Paul Smith's College of Arts and
 Sciences
Forestry Division
Paul Smiths, NY 12970

SUNY College of Environmental Science
and Forestry
School of Forestry, Forest Technician
Program
Wanakena, NY 13695

Haywood Community College
Agricultural and Biological Sciences Dept.
Freedlander Drive
Clyde, NC 28721

Hocking College
Natural Resource Department
Nelsonville, OH 45764

Eastern Oklahoma State College
Forestry Department
Wilburton, OK 74578

Central Oregon Community College
Forest Technology Program
N.W. College Way
Bend, OR 97701

Pennsylvania State University
Forest Technician Unit
Mont Alto Campus
Mont Alto, PA 17237

Pennsylvania College of Technology
Natural Resources Management
1005 W. Third Street
Williamsport, PA 17701

Horry-Georgetown Technical College
Forestry Department
P. O. Box 1966
Conway SC 29526

Spokane Community College
Natural Resources Department
N. 1810 Greene Street
Spokane, WA 99207

Dabney S. Lancaster Community College
Forest Technology Program
Route 60 West
Clifton Forge, VA 24422

Glenville State College
Department of Forest Technology
Glenville, WV 26351

Source: *Journal of Forestry*
March 1992, p. 26

Table 14
THE GROUP OF 10

Headquarters	Regional Offices	Representatives
DEFENDERS OF WILDLIFE		
1244 19th Street, NW	Missoula, MT	406-549-0761
Washington, DC 20036	Sacramento, CA	916-442-6386
202-659-9510		
ENVIRONMENTAL DEFENSE FUND		
257 Park Avenue South	Washington, DC	202-387-3500
New York, NY 10010	Oakland, CA	415-658-8008
212-505-2100	Boulder, CO	303-440-4901
	Richmond, VA	804-780-1297
	Raleigh, NC	919-821-7793
	Austin, TX	512-478-5161
IZAAK WALTON LEAGUE OF AMERICA, INC.		
1401 Wilson Blvd., Level B	Minneapolis, MN	612-922-1608
Arlington, VA 22209		
703-528-1818		
NATIONAL AUDUBON SOCIETY		
700 Broadway	Washington, DC	202-547-9009
New York, NY 10003-9501	Camp Hill, PA	717-763-4985
212-797-3000	Albany, NY	518-869-9731
	Waitsfield, VT	802-496-5727
	Dover-Foxcroft, ME	207-564-7946
	Tallahassee, FL	904-222-2473
	Columbus, OH	614-224-3303
	St. Paul, MN	612-225-1830
	Manhattan, KS	913-537-4385
	Austin, TX	512-327-1946
	Santa Fe, NM	505-983-4609
	Boulder, CO	303-499-0219
	Sacramento, CA	916-481-5332
	Olympia, WA	206-786-8020
	Anchorage, AK	907-276-7034
NATIONAL PARKS AND CONSERVATION ASSOCIATION		
1015 31st Street, NW	Tuscon, AZ	602-529-6232
Washington, DC 20007-4406	Salt Lake City, UT	801-532-4796
202-944-8530	Norris, TN	615-494-9786
	Des Moines, WA	206-824-8808
	Anchorage, AK	907-277-6722
	Albany, NY	518-434-1583

NATIONAL WILDLIFE FEDERATION

1400 16th Street, NW	Bismark, ND	701-222-2442
Washington, DC 20036-2266	Atlanta, GA	404-876-8733
202-797-6800	Ann Arbor, MI	313-769-3351
	Washington, DC	202-797-6693
	Montpelier, VT	802-229-0650
	Boulder, CO	303-492-6552
	Anchorage, AK	907-258-4800
	Missoula, MT	406-721-6705
	Portland, OR	503-222-1429

NATURAL RESOURCES DEFENSE COUNCIL, INC.

40 West 20th Street	Washington, DC	202-783-7800
New York, NY 10011	Honolulu, HI	808-533-1075
212-727-2700	San Francisco, CA	415-777-0220
	Los Angeles, CA	213-892-1500

SIERRA CLUB

730 Polk Street	Washington, DC	202-547-1141
San Francisco, CA 94109	Anchorage, AK	907-276-4048
415-776-2211	Los Angeles, CA	213-387-6528
	Madison, WI	608-257-4994
	Saratoga Springs, NY	518-587-9166
	Sheridan, WY	307-672-0425
	Seattle, WA	206-621-1696
	Phoenix, AZ	602-254-9330
	Birmingham, AL	205-933-9111
	Dallas, TX	214-369-8181
	Ottawa, Ontario	613-233-1906

SIERRA CLUB LEGAL DEFENSE FUND

180 Montgomery St., Suite 1400	Washington, DC	202-667-4500
San Francisco, CA 94104-4209	Seattle, WA	206-343-7340
415-627-6700	Denver, CO	303-623-9466
	Juneau, AK	907-586-2751
	Honolulu, HI	808-599-2436
	Tallahassee, FL	904-681-0031
	Bozeman, MT	406-586-9699
	New Orleans, LA	504-522-1394

THE WILDERNESS SOCIETY

900 17th Street, NW	Anchorage, AK	907-272-9453
Washington, DC 20006-2596	Boise, ID	208-343-8153
202-833-2300	San Francisco, CA	415-541-9144
	Atlanta, GA	404-872-9453
	Denver, CO	303-650-5818

Seattle, WA	206-624-6430
Portland, OR	503-248-0452
Boston, MA	617-350-8866
Augusta, ME	207-626-5635
Santa Fe, NM	505-986-8373
Coral Gables, FL	305-448-3636
Marathon, FL	305-289-1010
Bozeman, MT	406-586-1600

Source: Conservation Directory 1994
National Wildlife Federation

Table 15
NATIONAL WOODLAND OWNERS ASSOCIATION
FEDERATION OF WOODLAND OWNERS ASSOCIATIONS
STATE AFFILIATES

National Woodland Owners Association
374 Maple Avenue E., Suite 210
Vienna, VA 22180
703-255-2700
    Keith A. Argow, President

Alabama Forest Owners' Association
Box 104
Helena, AL 35080
205-663-1131
    R. Lee Laechelt

Forest Landowners of California
980 Ninth Street, Suite 1600
Sacramento, CA 95814
916-972-0273
    Daniel M. Weldon

Colorado Forestry Association
203 Forestry
CSU
Ft. Collins, CO 80523
303-491-6303
    Ruth Ann Steele

Connecticut Forest & Parks Association
16 Meriden Road, Route 66
Rockfall, CT 06481-2961
203-346-8733
    John E. Hibbard

Delaware Forestry Association
2320 South DuPont Highway
Dover, DE 19901
302-739-4811
    Sam Dyke

Idaho Forest Owners' Association
HC #1, Box 170
DeSmet, ID 83824
208-274-4573
    Joan Kertu

Illinois Woodland Owners and
    Users Association
RR 1, Box 57
Mason, IL 62443
618-245-6392
    Hilliard Morris

Indiana Forestry & Woodland
    Owners Association
2505 Radcliffe Avenue
Indianapolis, IN 46227
317-888-8484
    Mitchell Hassler

Iowa Woodland Owners Association
2404 S. Duff
Ames, IA 50011
515-233-1161
    William C. Ritter

Kentucky Woodland Owners Association
3385 Upper Tug Fork Road
Alexandria, KY 41001
606-635-7826
    Donald S. Girton

Small Woodland Owners of Maine
Box 926
Augusta, ME 04332-0926
207-626-0005
    Abbott Ladd

Massachusetts Forestry Association
Box 1096
Belchertown, MA 01007
413-323-7326
    Timothy Fowler

Michigan Forest Association
1558 Barrington Street
Ann Arbor, MI 48103
313-665-8279
    McLain Smith, Jr.

Minnesota Forestry Association
500 Exchange Bldg.
26 E. Exchange Street
St. Paul MN 55101
612-292-0051
    Douglas Ewald

New Hampshire Timberland
  Owners Association
54 Portsmouth Street
Concord, NH 03301
603-224-9699
    Charles Niebling

New Jersey Forestry Association
1628 Prospect St.
Trenton, NJ 08638
609-771-8301
    Ronald J. Sheay

The Forest Trust (New Mexico)
Box 519
Santa Fe, NM 87504-0519
505-983-8992
    Henry H. Carey

New York Forest Owners Association
Box 180
Fairport, NY 14450
716-377-6060
    John Marchant

Ohio Forestry Association
1301 Worthington Woods Blvd.
Worthington, OH 43085
614-846-9456
    Ronald C. Cornell

Oregon Small Woodlands Association
1149 Court Street NE
Salem, OR 97301
503-588-0050
    Gary M. Carlson

Pennsylvania Forestry Association
56 E. Main Street
Mechanicsburg, PA 17055
717-766-5371
    Patrick M. Lantz

Rhode Island Forest Conservators
Box 40328
Providence, RI 02908
401-273-8037
    Marc J. Tremblay

Vermont Timberland Owners Association
Box 72
E. St. Johnsbury, VT 05638
802-748-5560
    John Hemenway

Virginia Forestry Association
1205 E. Main Street
Richmond, VA 23219
804-644-8462
    Charles F. Finley, Jr.

Wash Farm Forestry Association
Box 7663
Olympia, WA 98507
206-459-0984
    Nels Hanson

Wisconsin Woodland Owners Association
Box 285
Stevens Point, WI 54481
715-341-4798
    Robert J. Engelhard

Source: NWOA
15 April 1994

Table 16
USDA, FOREST SERVICE, FIELD OFFICES

Northern Region (R-1)
Federal Building (Box 7669)
Missoula, MT 59807
406-329-3511

Rocky Mountain Region (R-2)
740 Simms Street (Box 25127)
Lakewood, CO 80401
303-275-5350

Southwestern Region (R-3)
Federal Building
517 Gold Avenue S.W.
Albuquerque, NM 87102
505-842-3292

Intermountain Region (R-4)
Federal Building
324 25th Street
Ogden, UT 84401
801-625-5352

Pacific Southwest Region (R-5)
630 Sansome Street
San Francisco, CA 94111
415-705-2874

Pacific Northwest Region (R-6)
333 S. W. 1st Avenue (Box 3623)
Portland, OR 97208
503-326-2971

Southern Region (R-8)
1720 Peachtree Road N.W.
Atlanta, GA 30367
404-347-2384

Eastern Region (R-9)
310 W. Wisconsin Avenue, #500
Milwaukee, WI 53203
414-297-3693

Alaska Region (R-10)
Box 21628
Juneau, AK 99802-1628
907-586-8863

Northeastern Area - S&PF
5 Radnor Corporate Center, Ste. 200
100 Matsonford Road (Box 6775)
Radnor, PA 19087-4585
215-975-4111

Intermount. Forest & Range Exp. Station
Federal Building
324 25th Street
Ogden, UT 84401
801-625-5412

N. Central Forest Exp. Station
1992 Folwell Avenue
St. Paul, MN 55108
612-649-5000

Northeastern Forest Exp. Station
5 Radnor Corporate Center, Ste.
100 Matsonford Rd. (Box 6775)
Radnor, PA 19087-4585
215-975-4222

Pacific NW Forest & Range Exp. Station
333 S. W. 1st Avenue (Box 3890)
Portland, OR 97208
503-326-3592

Pacific SW Forest & Range Exp. Station
800 Buchanan Street (Box 245)
Berkeley, CA 94701-0245
510-559-6300

Rocky Mt. Forest & Range Exp. Station
240 W. Prospect Road
Fort Collins, CO 80526-2098
303-498-1100

Southeastern Forest Exp. Station
200 Weaver Blvd. (Box 2680)
Asheville, NC 28802
704-257-4832

Southern Forest Exp. Station
US Postal Service Bldg., Room T-10210
701 Loyola Avenue
New Orleans, LA 70113
504-589-6800

Forest Products Laboratory
One Gifford Pinchot Drive
Madison, WI 53705-2398
608-231-9200

Source: USDA, Forest Service Organizational Directory (FS-65), June 1993

Table 17
NATIONAL ASSOCIATION OF STATE FORESTERS

Tom Boutlin
Division of Forestry
Box 107005
Anchorage, AK 99510-7005
907-561-6659

Timothy C. Boyce
Alabama Forestry Commission
513 Madison Avenue
Montgomery, AL 36130
205-240-9304

John T. Shannon
Arkansas Forestry Commission
3821 W. Roosevelt Road
Little Rock, AR 72214
501-664-2531

T. Michael Hart
State Land Department
1616 W. Adams
Phoenix, AZ 85007
602-542-2515

Richard Wilson
Department of Forestry
Resources Building
Box 94246
Sacramento, CA 94244-2460
916-653-7772

James E. Hubbard
Colorado State Forest Service
CSU, 203 Forestry Bldg.
Fort Collins, Co 80523
303-491-6303

Donald H. Smith
Bureau of Forestry
165 Capital Avenue
Hartford, CT 06106
203-566-5348

Acting State Forester
Forestry Section
2320 S. DuPont Highway

Dover, DE 19901
302-697-6287

Sandra Hill
Trees & Lands
District Government
2750 South Capitol St., S.E.
Washington, D.C. 20032
202-767-8968

Earl Peterson
Division of Forestry
3125 Conner Boulevard
Tallahassee, FL 32399-1650
904-488-0863

John W. Mixon
Georgia Forestry Commission
Box 819
Macon, GA 31298-4599
912-751-3465

Carlos L.T. Noquez
Forestry & Soil Resources Division
Box 2950
Agana, Guam 96910
671-734-3948

Michael Buck
Division of Forestry & Wildlife
1151 Punchbowl Street
Honolulu, HI 96813
808-587-0166

William Farris
Department of Natural Resources
Wallace Office Bldg., E. 9th & Grand
Des Moines, IA 50319
515-281-8656

Stanley F. Hamilton
Idaho Department of Lands
1215 West State Street
Boise, ID 83720-7000
208-334-0200

Stewart Pequingnot
Division of Forest Resources
600 North Grand Avenue West
Springfield, IL 62706

Burnell C. Fischer
Department of Natural Resources
402 W. Washington Street, #269
Indianapolis, IN 46204
317-232-4105

Raymond Aslin
State & Extension Forestry
2610 Claflin Road
Manhattan, KS 66502
913-537-7050

Mark Matuszewski
Kentucky Division of Forestry
627 Comanche Trail
Frankfort, KY 40601
502-564-4496

Paul D. Frey
Office of Forestry
Box 1628
Baton Rouge, LA 70821
504-925-4500

Warren Archey
Division of Forests and Parks
100 Cambridge Street
Boston, MA 02202
617-727-3180, Ext. 685

John Riley
Forest, Park & Wildlife Service
Tawes State Office Building
580 Taylor Avenue
Annapolis, MD 21401
410-974-3776

Susan J. Bell
Maine Forest Service
State House Station #22
Augusta, ME 04333
207-287-2793

Gerald Thiede
Forest Management Division

Michigan Dept. of Natural Resources
Box 30028
Lansing, MI   48909
517-373-1275

Gerald Rose
Division of Forestry
500 Lafayette Road
St. Paul, MN 55155-4044
612-296-4484

Marvin Brown
Missouri Dept. of Conservation
Box 180
Jefferson City, MO 65102-01801
314-751-4115

James Sledge
Mississippi Forestry Commission
301 Building, Suite 300
Jackson, MS 39201
601-359-1389

Donald K. Artley
Division of Forestry
2705 Spurgin Road
Missoula, MT 59801
406-542-4300

Stan Adams
Division of Forest Resources
Box 27687
Raleigh, NC 27611-7687
919-733-2162

Larry A. Kotchman
North Dakota Forest Service
First and Brander
Bottineau, ND 58318
701-228-2277

Gary L. Hergenrader
Dept. of Forestry, Fish & Wildlife
Plant Industries Bldg, Room 101
Lincoln, NE 68583
402-472-2964

John E. Sargent
Division of Forests & Lands
Box 856

Concord, NH 03302-0856
603-271-2214

Olin D. White, Jr.
New Jersey Forest Servic
CN 404, 501 E. State Street
Station Plaza #5
Trenton, NJ 08625
609-984-3850

Raymond R. Gallegos
Forestry and Resource Conservation
Division
Box 1948
Santa Fe, NM 87504-1948
505-827-5830

Roy Trenoweth
Division of Forestry
123 W. Nye Lane, Suite 142
Carson City, NV 89710
702-687-4353

Robert H. Bathrick
Division of Lands & Forests
50 Wolf Road
Albany, NY 12233
518-457-2475

Ronald Abraham
Division of Forestry
4383 Fountain Square
Columus, Oh 43224
614-265-6690

Roger L. Davis
Oklahoma Department of Agriculture
Forestry Services
2800 North Lincoln Boulevard
Oklahoma City, OK 73105
405-521-3864

James E. Brown
Oregon Department of Forestry
2600 State Street
Salem, OR 97310
503-945-7211

James R. Grace
Bureau of Forestry

Box 8552
Harrisburg, PA 17105-8552
717-787-2703

Jorge Palou
Forest Service
Department of Natural Resources
Box 5887
Puerta de Tierra
San Juan, PR 00906
809-724-3647

Thomas A. Dupree
Division of Forest Environment
1037 Hartford Pike
North Scituate, RI 02857
401-647-3367

Hugh Ryan
South Carolina Forestry Commission
Box 21707
Columbia, SC 29221
803-896-8800

Frank Davis
Division of Forestry
Sigurd Anderson Building
445 East Capitol Ave.
Pierre, SD 57501
605-773-3623

Kendrick S. Arney
TN Dept. of Agr, Div of Forestry
Box 40627, Melrose Station
Nashville, TN 37204
615-360-0722

Bruce Miles
Texas Forest Service
College Station, TX 77843-2136
409-845-2601

Robert E. Storey
Department of Natural Resources
3 Triad Center, Suite 400
Salt Lake City, UT 84180-1204
801-538-5508

James W. Garner
Virginia Division of Forestry

Box 3758
Charlottesville, VA 22903
804-977-6555

Eric L. Bough
Virgin Island Forestry Program
King Field Post Office
St. Croix, VI 00850
809-778-0097, Ext. 19

Conrad M. Motyka
Department of Forests, Parks and Recreation
103 S. Main Street, 10 South
Waterbury, VT 05676
802-241-3670

Kathleen Cottingham
Department of Natural Resources
Box 47001
Olympia, WA 98504-7001
206-902-1000

Charles Higgs
Department of Natural Resources
Box 7921
Madison, WI 53707
608-266-0842

William R. Maxey
Forestry Division
1900 Kanawha Blvd., East
Charleston, WV 25305-0180
304-558-3446

Michael Gagen
Wyoming State Forestry Division
1100 West 22nd Street
Cheyenne, WY 82002
307-637-8726

Washington Representative:
Terry Bates
444 N. Capitol St., Suite 540
Washington, D.C. 20001
202-624-5415

Source: National Association of State Foresters
Washington, DC, March 1994

Table 18
NATIONAL COUNCIL OF FORESTRY ASSOCIATION EXECUTIVES

Alabama Forestry Association
555 Alabama Street
Montgomery, AL 36104
John McMillan, Executive Vice President
205-265-8733

Alaska Forestry Association
111 Stedman, Suite 200
Ketchican, AK 99901
Larry Blasing
907-225-6114

Arkansas Forestry Association
410 S. Cross Street, #C
Little Rock, AR 72201
Chris Barneycastle
501-374-2441

Associated California Loggers
555 Capitol Mall, #745
Sacramento, CA 95814
Ed Ehlers
916-441-7940

California Forestry Association
1311 I Street, Suite 100
Sacramento, CA 95814
William Dennison, President
916-444-6592

Connecticut Forest & Parks Association, Inc.
16 Meriden Road, Route 66
Rockfall, CT 06481-2961
John Hibbard, President
203-346-8733

Empire State Forest Products Association
123 State Street
Albany, NY 12207
Kevin King, Executive Vice President
518-463-1297

Florida Forestry Association
Box 1696
Tallahassee, FL 32302

Wm. Carroll Lamb, Executive Vice President
904-222-5646

Georgia Forestry Association
505 Pinnacle Court
Norcross, GA 30071
Robert L. Izlar, Executive Director
404-416-7621

Indiana Forestry and Woodland Owners
Association
2505 Radcliffe Avenue
Indianapolis, IN 46227
Mitchell Hassler, Executive Director
317-888-8484

Kentucky Forest Industries Association
310 Kings Daughter Drive, #7
Frankfort, KY 40601
James H. Lee, Executive Director
502-875-3136

Louisiana Forestry Association
Drawer 5067
Alexandria, LA 71307-5067
Charles A. Vandersteen, Executive Director
318-443-2558

Maine Forest Products Council
146 State Street
Augusta, ME 04332
Edward I. Johnson, Executive Director
207-622-9288

Michigan Association of Timbermen
PO box 486
Newberry, MI 49868
Peter Grieves, Executive Director
906-293-3236

Minnesota Forestry Association
500 Exchange Building
26 E. Exchange Street
St. Paul, MN 55101
Cristy Muller Holden, Executive Director
612-292-0051

Mississippi Forestry Association
620 North State Street, #201
Jackson, MS 39202-3398
Steve A. Corbitt, Executive Vice President
601-354-4936

Missouri Forest Products Association
611 E. Capitol Avenue, Suite 1
Jefferson City, MO 65101
Ken Christgen, Executive Director
314-634-3252

Montana Logging Association
PO Box 1716
Kalispell, MT 59903-1716
Keith L. Olson, Executive Director
406-752-3168

Montana Wood Products Association
208 N. Montana Avenue, Suite 104
Helena, MT 59601
Don L. Allen, Executive Vice President
406-443-1566

New Hampshire Timberland Owners
Association
54 Portsmouth Street
Concord, NH 03301
Charles Niebling, Executive Director
603-224-9699

Society for the Protection of New Hampshire Forests
54 Portsmouth Street
Concord, NH 03301-5400
Paul O. Bofinger, President
603-224-9945

New Jersey Forestry Association
1628 Prospect Street
Trenton, NJ 08638
Ronald J. Sheay
609-984-3866

New York State Timber Producers
Box 300
Boonville, NY 13309
John R. Courtney. Jr., President
315-942-5503

North Carolina Forestry Association
Box 12825
Raleigh, NC 27605
Robert W. Slocum, Jr., Executive Vice
President
919-834-3943

Ohio Forestry Association
1301 Worthington Woods Boulevard
Worthington, OH 43085
Ronald Cornell, Executive Vice President
614-846-9456

Oklahoma Forestry Association
Box 517
Broken Bow, OK 74728
Dale Campbell, Executive Vice President
405-584-6911

Pennsylvania Forestry Association
56 E. Main Street
Mechanicsburg, PA 17055
Patrick Lantz, President
717-766-5371

Small Woodlot Owners Association of Maine
PO Box 926
Augusta, ME 04332
Abbott Ladd, Executive Director
207-626-0005

South Carolina Forestry Association
Box 21303
Columbia, SC 29221
Robert R. Scott, President
803-798-4170

Tennessee Forestry Association
PO Box 290693
Nashville, TN 37229
Candace Dinwiddie
615-883-3832

Texas Forestry Association
PO Box 1488
Lufkin, TX 75902-1488
Ronald H. Hufford, Executive Vice President
409-632-TREE

Virginia Forestry Association
1205 E. Main Street
Richmond, VA 23219
Charles F. Finley, Jr., Executive Vice President
804-644-8462

West Virginia Forestry Association
PO Box 724
Ripley, WV 25271
Dick Waybright, Executive Director
304-273-8164

Washington Forest Protection Association
711 Capitol Way, Room 608
Olympia, WA 98501
William C. Jacobs, Executive Director
206-352-1500

Wisconsin Woodland Owners Association
PO Box 285
Stevens Point, WI 54481
Robert J. Engelhard, Executive Director
715-341-4798

Source: National Council of Forestry Association Executives
29 April 1992

ENDNOTES

CHAPTER 1:
FORESTLAND ECOSYSTEMS: WHERE TO FROM HERE

1. AFPA Report, American Forest and Paper Association, Washington, DC, July 1993.
2. Senator Al Gore, *Earth in the Balance* (New York, NY: Houghton Miflin Co., 1992), p. 119.
3. William M. Harlow, Ph.D. and Ellwood S. Harrar, Ph.D., *Textbook of Dendrology* (New York: McGraw-Hill Book Co, Inc., 1941), p. 234.

CHAPTER 2:
FORESTLAND ECOSYSTEMS: PANDORA'S BOX

1. Theodore Roosevelt, *An Autobiography* (New York, NY: Charles Scribner's Sons, 1923), p. 401.
2. Richard C. Davis, ed., *Encyclopedia of American Forest and Conservation History* (New York, NY: Macmillan Publishing Co., 1983), for this and following pertinent quotes.
3. "The Principal Laws Relating to Forest Service Activities," USDA, Forest Service, 1993.
4. *Report of the Forest Service, Fiscal Year 1992*, United States Department of Agriculture, Forest Service, Washington, DC, February 1993, p. 5.
5. *Statistical Abstract of the United States*, 1991, U.S. Department of Commerce, Bureau of the Census, p. 203, Table #350.
6. "Biologist Has Ants in Pants," article on Dr. Edward O. Wilson, *Montgomery Advertiser*, 29 March 1993.
7. Christopher Lehmann-Haupt, "Finding the Beauty and the Otherness of Bugs," *New York Times*, 24 June 1993, p. C18; book review, Sue Hubbell, *Broadsides from the Other Side*, Random House.
8. Davis, p. 694.
9. William B. Greeley, *Forests and Men* (Garden City, NY: Doubleday & Co., Inc., 1956), p. 66.
10. *The Forest, Environmental Protection in Bavaria*, Bavarian Ministry for Food, Agriculture and Forest, 1989, p. 21.
11. Andy Stahl, Sierra Club Legal Defense Fund, Sixth Annual Western Public Interest Law Conference, University of Oregon Law School, 5 March 1988.
12. *Barron's*, Dow Jones & Company, Inc., New York, NY, undated.
13. NBC news, "Today," 26 April 1989.
14. ABC "World News," 26 April 1989.
15. Jim MacNeil, "MacNeil/Lehrer News Hour," undated.
16. "Ancient Forests: Rage Over Trees": National Audubon Society, PBS television program, 26 August 1991, Paul Newman, narrator.
17. Land and Resource Management Plan, Willamette National Forest, USDA Forest Service, Pacific Northwest region, 1990, Table G-1, Appendix G; Mike Morris, phone interview 3 October 1991.
18. Dr. William A. Atkinson, "Old Forestry in a New World," presentation to Western Forestry Conference, Coeur d'Alene, ID, 3 December 1990, p. 12, Figure 9.
19. Rob Taylor, "Reluctant Leader of Forest Changes Leaves a Warning," Seattle P-I, 2 October 1992, p. B2.
20. Nova, "Return to Mt. St. Helens," shown several times on Public Television.

CHAPTER 3:
STEWARDSHIP

1. Ellis Lucia, *Head Rig* (Portland, Oregon: Overland West Press, 1965), p. 220.
2. William B. Greeley, *Forests and Men* (Garden City, N.Y: Doubleday & Co., 1956), p. 18.
3. Ibid., p. 18.
4. Gerald J. Gray, ed., *Resource Hotline*, American Forestry Association, Vol. 7, No. 18, p. 2.
5. Greeley, p. 202.
6. Ibid., p. 201.
7. I am grateful to Kathy Porter, assistant editor, for her coverage of the Bear Story, "An Evening with Bruce Vincent," Wood Tick Trail, *Timber Harvesting*, August 1991. Also phone calls to Patty Jo Vincent, 8 October 1991; to Bruce Vincent 26 December 1991 and 7 January 1992.
8. Porter, p. 7.
9. Ibid.
10. Ibid., p. 6.
11. *Fortune*, Time Inc. Magazine Co., New York, 15 July 1966.
12. Ken Wells, "The Battles Are Over, But Gulf Environment Still Fights for Its Life," *Wall Street Journal*, 15 October 1991, p.1.
13. Ibid.
14. Paid advertisement, *New York Times*, 6 February 1990, p. C7 and 24 March 1994, p. B9.
15. Douglas H. Strong, "The Sierra Club—A History," Sierra Club Public Affairs, San Francisco, CA, reprint from *Sierra*, October and November/December 1977, p. 3.
16. Ibid., p. 3
17. Ibid., p. 2.
18. Ibid., p.7.
19. Ibid.
20. Ibid.
21. *New York Times*, 6 February 1990, p. C7.
22. "California Petitions to Delist Owl," *Green Speak*, November 1993.
23. "Red-Taped Housing," *Wall Street Journal*, 4 September 1991, editorial page.
24. *WSJ*, 4 September 1991, for this and following report quotes.
25. Mitchell Pacelle, "More Builders Plan Low-Cost, Small Houses," *Wall Street Journal*, 30 August 1991, p. B1.
26. By 1990, there were a total of 21.1 million jobs; AFPA Forest Resources, Washington, DC, 7 June 1992.
27. Ellis Lucia, *Head Rig* (Portland, OR: Overland West Press, 1965), p. 209.
28. *Random Lengths*, Random Length Publications, Inc., Eugene, OR, 20 March 1992, 19 March 1993.
29. V. Alaric Sample, *Land Stewardship in the Next Era of Conservation*, Pinchot Institute for Conservation, Grey Towers National Historic Landmark, Milford, PA, 1991, p. ix.
30. George Perkins Marsh, *The Earth as Modified by Human Action* (New York, NY: Scribner, Armstrong & Co., 1874), p. 343.
31. Ibid., p. 344.
32. HMA, unpublished research data, Weyerhaeuser Company.
33. Charles E. Twining, *Phil Weyerhaeuser: Lumberman* (Seattle, WA: University of Washington Press, 1985), p. 15.
34. *A Celebration for Generations to Come, Fifty Years, Clemons Tree Farm, 1941-1991*,

Weyerhaeuser Company, Tacoma, Washington, 1990, pp. 5, 7.
35. *Resource Stewardship at Willamette Industries*, Willamette Industries, Inc., 1989.
36. "Wildlife & Westvaco at Wickliffe," Westvaco publication.
37. R. Scott Wallinger, Westvaco News Release, 20 June 1991.

CHAPTER 4:
NEGATIVE AND POSITIVE CASE HISTORIES

1. David M. Smith, *The Practice of Silviculture* (New York: NY: John Wiley & Sons, 1986), eighth edition.
2. "AFA Today," *American Forests*, January/February 1991, p.12.
3. Smith, Preface, vii.
4. Scott Berg, "Forestry's Role in Wetlands Conservation," *Tree Farmer*, Summer 1990, p. 11.
5. "Capital Communique," Southern Forest Products Association, 14 March 1994.
6. Ibid.
7. Mitch Dubensky, "Wetlands Regulations—Beware the Net," *Tree Farmer*, Summer 1990, p. 12.
8. Mike Tankersley, "The Issue of Wetlands," *Timber Harvesting*, June 1991, p. 43: Dubensky, p. 12.
9. Tankersley, p. 43.
10. Gary Moll, *Urban Forests*, American Forest Association, Washington, DC, December/January 1992, p. 3, 5, 11, and 12.
11. Peter Drucker, "It Profits Us to Strengthen Nonprofits," *Wall Street Journal*, 19 December 1991.

CHAPTER 5:
TREES AND CORPORATIONS

1. *Tappi Journal*, September 1992, p. 16.
2. "Investor's Guide," Fortune, Autumn 1993, p. 8.
3. Report RM-199, p. 250.
4. Peter F. Drucker, *Practice of Management* (New York, N.Y.: Harpers & Brothers, Inc., 1954), p. 37.
5. Therese Eiben and John Labate, "How the Industries Stack Up," *Fortune*, 11 January 1993, pp. 66-73.
6. Alston Chase, *Playing God in Yellowstone* (Boston/New York: Atlantic Monthly Press, 1986), p. 240.
7. Ibid., p. 241.
8. Ibid., p. 32.

CHAPTER 6:
A CONSERVATION ETHIC FOR TREES AND FORESTLANDS

1. Samuel Trask Dana, *Forest and Range Policy, Its Development in the United States* (New York, NY: McGraw-Hill Book Co, Inc., 1956), p. 3.
2. Ibid., p. 5.
3. Richard H. Stroud, ed., *National Leaders of American Conservation* (Washington, DC: Smithsonian Institution Press, 1985), p. 97.

4. Henry Clepper, *Crusade for Conservation* (Washington, DC: American Forestry Association, 1975).
5. Clepper, p. 85; Stroud, p. 397; Andrew Denny Rodgers III, *Bernard Eduard Fernow, A Story of North American Forestry* (Durham, NC: Forest History Society, 1991), p. 48.
6. Clepper, *Crusade*, p. 11.
7. Ibid.
8. William N. Sparkhawk, "History of Forestry in America," *Trees: The Yearbook of Agriculture*, 1949, U.S. Department of Agriculture, Washington, DC, p. 705.
9. Davis, p. 528; Stroud, p. 306.
10. Clepper, *Crusades*, p. 16; Davis, p. 297; Stroud, p. 209.
11. Clepper, *Crusades*, p. 16.
12. Ibid., p. 17.
13. Ibid., 12.
14. Ibid., p. 13.
15. Ibid., p. 13; Davis, p. 168; Stroud, p. 149.
16. Clepper, *Crusades*, p. 13.
17. Ibid., p. 13.
18. Ibid., 16.
19. Ibid.
20. Davis, p. 405.
21. George Perkins Marsh, *The Earth as Modified by Human Action* (New York, NY: Scribner, Armstrong & Co., 1874), p. 368.

CHAPTER 7:
TREES AND EDUCATION

1. Henry Clepper, *Crusade for Conservation*, Washington, DC, 1975, p. 3.
2. Richard C. Davis, ed, *Encyclopedia of American Forest and Conservation History* (New York: Macmillan Publishing Co., 1983), p. 298.
3. Henry Clepper, *Professional Forestry in the United States* (Baltimore and London: Johns Hopkins Press, 1971).
4. Carl Alwin Schenck, *The Birth of Forestry in America, Biltmore Forest School, 1898-1913*, Forest History Society and the Appalachian Consortium, Santa Cruz, California, 1974, p. 8, 13.
5. Davis, p. 237.
6. R. Scott Wallinger, "Creating and Educating a 21st Century Forest Resources Profession," Plenary Session Presentation, *Symposium Proceedings, Forest Resource Management in the 21st Century*, Oct.30-Nov.2, 1991, Denver, Colorado, pp. 30-38.

CHAPTER 8:
TREES AND GOVERNMENT

1. Henry Clepper, *Crusade for Conservation* (Washington, DC: American Forestry Association, 1975), p. 18.
2. Ibid., p. 19.
3. Ibid.
4. Ibid.
5. Ibid., p. 20.

6. Ibid.
7. Ibid.
8. William B. Greeley, *Forests and Men* (Garden City, N.Y.: Doubleday & Company, Inc., 1956), p. 59.
9. Clepper, *Crusade*, p. 21.
10. Ibid.
11. American Forestry Association, *Proceedings of the American Forest Congress 1905* (Washington, DC: H. M. Publishing Co., 1905), Preface.
12. Clepper, *Crusade*, pp. 22, 23.
13. Henry Clepper, "Year of Destiny," *American Forests*, October 1980, p. 37.
14. Samuel Trask Dana, *Forest and Range Policy, Its Development in the United States* (New York, N.Y.: McGraw-Hill Book Company, 1957, p. 119.
15. *The Economist*, 10 March 1990.
16. Henry Clepper, *Professional Forestry in the United States* (Baltimore and London: Johns Hopkins Press, 1971), p. 29.
17. Lenore K. Bradley, *Robert Alexander Long, A Lumberman of the Guilded Age*, Forest History Society, Durham, NC, 1989, p. 57.
18. Clepper, *Professional Forestry*, p. 45.
19. Ibid., p. 46.
20. Ibid.
21. George T. Morgan, Jr., *William B. Greeley, A Practical Forester*, Forest Historical Society, 1961, p. 2, 24.
22. Clepper, *Crusade*, p. 28.
23. Ibid., pp. 28, 29.
24. Bradley, p. 57.
25. Ibid., p. 59.
26. Morgan, p. 33.
27. Ibid.
28. Ibid.
29. Morgan, p. 37.
30. Ibid., p. 39.
31. Ibid.
32. Ibid., p. 54.
33. Greeley, p. 112.
34. *Report of the Forest Service, Fiscal Year 1992*, United States Department of Agriculture, Forest Service, Washington, DC, February 1993, p. 73.

CHAPTER 9:
HAVE YOU CARED FOR A TREE TODAY?

1. Robert Weir, "Forest for Sale: It's a Steal," *New York Times*, 4 March 1992.
2. Chuck Dolce, Letters, *Earthwatch Magazine*, June 1991, p. 2.
3. John Muir, *The Mountains of California* (New York, NY: Dorset Press, 1988— originally pub. in 1894), p. 152.

INDEX